MODERN DIESEL TECHNOLOGY: MOBILE EQUIPMENT HYDRAULICS: A SYSTEMS AND TROUBLESHOOTING APPROACH

Ben Watson

SAE, ASE

Australia • Brazil • Japan • Korea • Mexico • Singapore • Spain • United Kingdom • United States

Modern Diesel Technology:
Mobile Equipment Hydraulics:
A Systems and Troubleshooting Approach
Ben Watson

Vice President, Editorial: Dave Garza

Director of Learning Solutions: Sandy Clark

Executive Editor: David Boelio

Managing Editor: Larry Main

Senior Product Manager: Sharon Chambliss

Editorial Assistant: Jillian Borden

Vice President, Marketing: Jennifer Baker

Executive Marketing Manager: Deborah S. Yarnell

Marketing Specialist: Mark Pierro

Production Director: Wendy Troeger

Production Manager: Mark Bernard

Content Project Manager: Cheri Plasse

Art Director: Benj Gleeksman

Technology Project Manager: Chrstopher Catalina

Production Technology Analyst: Thomas Stover

For product information and technology assistance, contact us at
Professional & Career Group Customer Support, 1-800-648-7450

For permission to use material from this text or product, submit all requests online at **cengage.com/permissions**. Further permissions questions can be e-mailed to **permissionrequest@cengage.com**.

Library of Congress Control Number: 2010926037

ISBN-13: 978-1-4180-8043-3
ISBN-10: 1-4180-8043-8

Delmar
Executive Woods
5 Maxwell Drive
Clifton Park, NY 12065-2919
USA

Cengage Learning products are represented in Canada by Nelson Education, Ltd.

To learn more about Delmar, visit **www.cengage.com/delmar**.

Purchase any of our products at your local college store or at our preferred online store **www.cengagebrain.com**.

Notice to the Reader
Publisher does not warrant or guarantee any of the products described herein or perform any independent analysis in connection with any of the product information contained herein. Publisher does not assume, and expressly disclaims, any obligation to obtain and include information other than that provided to it by the manufacturer. The reader is expressly warned to consider and adopt all safety precautions that might be indicated by the activities described herein and to avoid all potential hazards. By following the instructions contained herein, the reader willingly assumes all risks in connection with such instructions. The publisher makes no representations or warranties of any kind, including but not limited to, the warranties of fitness for particular purpose or merchantability, nor are any such representations implied with respect to the material set forth herein, and the publisher takes no responsibility with respect to such material. The publisher shall not be liable for any special, consequential, or exemplary damages resulting, in whole or part, from the readers' use of, or reliance upon, this material.

Printed in the USA
2 3 4 5 6   27 26 25 24 23

Brief Contents

Contents

Preface

I have been teaching hydraulics and fluid power systems to motor vehicle and heavy equipment technicians for over 20 years. Most of the students I have had in classes were journeymen vehicle service technicians. These technicians generally do diagnostics and repair on every system, from the front bumper to the rear cross member. In most cases it was evident that the initial professional training they received prepared them well for their careers. In the case of fluid power systems, it was equally evident that many of these journeymen doubted their ability to troubleshoot and repair.

For over a decade of my career, I traveled extensively, conducting training across the North American continent and into the Pacific Rim. During this time I had the opportunity to meet many diesel- and heavy-equipment instructors. In talking to them, a common thread seemed to be that although there were many quality books on fluid power systems, they were generally geared toward the technician intending to make a career in maintaining stationary and mobile hydraulic systems, and not for those interested in maintaining trucks and heavy equipment that included hydraulic systems. These books frequently left these instructors with the impression that they were intended for engineers designing hydraulic systems and not for truck technicians.

In researching and planning this book I tried to balance the perspective I had as a novice technician with that of a professional trainer and educator. As a novice technician, I was more interested in knowing how to diagnose and repair than I was in knowing the "science" of a system. As a more mature technician and as an educator, I know that understanding the science of a technology is at the very core of diagnosis. This is especially true when the symptoms are not commonplace and the usual list of faults do not prove to be the problem. There are many books specific to the maintenance of aircrafts, stationary equipment, and to equipment of a particular manufacturer. I felt the industry needed a book that would address fluid power systems, also called hydraulic systems, from the perspective of what the truck service technician needed to know to maintain and repair truck-mounted equipment. This book is therefore a result of the lack of learning literature that was available for the mobile equipment technician learning his craft. In this book, I have tried to avoid the use of formal language and instead use technicians' language, and approach each topic from a technician's perspective whenever possible.

When teaching, the best way to organize a course is to identify the knowledge or skill level of the student and move them forward, step-by-step to the targeted skill or knowledge level. Each programmed learning step should build on those skills that have already been achieved or mastered.

Most people destined for a career as a technician are directed toward that career because on some level they are disillusioned with the standard academic path. I remember asking most of my math teachers to give me an example of how I could apply an exciting new formula to everyday life or to something I might do in the future on a job. Only one math teacher in my primary and secondary education was able to do that. She showed me how a math formula could be used to determine if gas had leaked from a pressurized cylinder or if a temperature change was the cause of a pressure change. There is no doubt that this experience in my junior year of high school was the moment that confirmed my education up to that point had meaning.

When planning this book, the intent was to take the student quickly into a practical overview and a high level understanding of how a basic fluid power system works. The intent is to establish interest and enthusiasm quickly. Once that is achieved, the book takes the student through the finer level details of materials, controls, circuit design, and fluids.

ORGANIZATION

The layout of this book permits instructors with very little classroom time to devote to fluid power to concentrate their instruction on a single chapter, **Chapter 2**.

This chapter is a focal point that can be used to help the students attain a general understanding of how a basic fluid power system operates. The subsequent chapters can then be made self-study assignments. For instance, if the program curriculum content concentrates mostly on line haul trucks, then **Chapter 2** can be covered in depth and a student made aware that additional information they may need later in their career is available in the rest of the book. If the program curriculum content is oriented toward vocational trucks and heavy equipment, then **Chapter 2** can be used as a focal point to which all subsequent chapters can be referenced.

Spread throughout the chapters, and in the appendices, are numerous tables and charts that can be helpful throughout a technician's career. These can also be used to help instructors develop exercises and tests to enhance the competency and skill level of students.

INTENDED AUDIENCE

This book is designed specifically to meet the needs of the motor vehicle service technician student. The design of the book recognizes the fact that for most "truck mechanics," making diagnoses and repairing fluid power systems will be only a small part of their overall job experience. The intention is to familiarize students with fluid power systems so that they can do research and make diagnostic decisions based on a foundational understanding of the system and an understanding of where and how to find essential information about the system.

ABOUT THE SERIES

The Modern Diesel Technology (MDT) series has been developed to address a need for *modern*, system-specific text-books in the field of truck and heavy equipment technology. This focused approach gives schools more flexibility in designing programs that target specific ASE certifications. Because each text-book in the series focuses exclusively on the competencies identified by its title, the series is an ideal review and study vehicle for technicians prepping for certification examinations.

Titles in the Modern Diesel Technology Series include:

MDT: Electricity and Electronics, by Joe Bell; ISBN: 1401880134

MDT: Heating, Ventilation, Air Conditioning, and Refrigeration, by John Dixon; ISBN: 1401878490

MDT: Electronic Diesel Engine Diagnosis, by Sean Bennett; ISBN: 1401870791

MDT: Brakes, Suspension, and Steering Systems, by Sean Bennett; ISBN: 1418013722

MDT: Heavy Equipment Systems, by Robert Huzij, Angelo Spano, Sean Bennett, and George Parsons; ISBN: 1418009504

MDT: Preventive Maintenance and Inspection, by John Dixon; ISBN: 1418053910

MDT: Mobile Equipment Hydraulics: A Systems and Troubleshooting Approach; ISBN: 1418080438

ACKNOWLEDGMENTS

I would like to thank three individuals who have greatly influenced my career as a technician, instructor, and writer.

I would like to thank Mr. C.R. Byford, who took me under his wing during my young technician era. C.R. taught me that the most important information in troubleshooting always comes from the equipment operator. "Talk to the operator long enough, ask the right questions, listen to the truths behind the answer, and he/she will tell you exactly where the problem is located." Too many technicians develop arrogance about the trade and begin to assume that the operator who is not an expert technician has nothing of value to say.

I would like to thank Dr. Chapin Ross, who taught me that to prepare a student for their first job in an industry is of little use. The world is full of people who can perform adequately in an entry-level capacity. To truly benefit a student, one must lay out a path of knowledge that leads to an ability to perform in the terminal capacity of that profession. If teaching accounting, the instructor should develop curriculum that prepares the student to be CFO (chief financial officer) not just a bookkeeper. In preparing a technician, the curriculum should be designed to prepare not just a functioning apprentice, but a journeyman technician who is capable transferring his or her skills, abilities, and experiences to the next generation of technicians. Understanding the *important* science of a technology is key to this.

Finally I would like to thank Benjamin Watson, my father. He taught me that an understanding of the fundamental principles of physics, mathematics, and chemistry is essential to performing effectively at any level in a profession. He taught me to apply skills learned in one technology to other technologies. He taught me that the "jack of all trades, master of none" is far more valuable in the service sector professions than is the dedicated technology specialist, the master of a single skill.

An additional thanks needs to go out to the dozens of fellow instructors, educators, and technicians who have loaned me their time, knowledge, opinion, and expertise in the development of this book.

The author and publisher would like to thank the following individuals for their comments and suggestions during the development process:

Ronald Scoville, J. Sargeant Reynolds Community College, Richmond, Virginia

Robert Huzij, Cambrian College, Sudbury, Ontario Canada

SUPPLEMENT

An Instructor Resources CD is available with the textbook. Components of the CD include an electronic copy of the Instructor's Guide, PowerPoint® lecture slides that present the highlights of each chapter, and an Image Gallery that includes an electronic copy of the images in the book.

CHAPTER

An Overview

Learning Objectives

Upon completion and review of this chapter, the student should be able to:

- Describe fluid flow through a typical cylinder-based hydraulic system.
- Describe fluid flow through a typical motor-based hydraulic system.
- Predict how pressure changes as fluid flows through a typical hydraulic system.
- Predict how temperature changes as fluid flows through a typical hydraulic system.
- Predict which components in a typical hydraulic system can affect flow.
- Predict where and how velocity will change in a typical hydraulic system.
- Describe how changes in pressure cause changes in force.
- Describe how changes in flow rate cause changes in the speed of operation in a system.

Cautions for This Chapter

- When inspecting a system, remember that the hydraulic fluid may be under high pressure.
- When working around a system with leaks, remember that hydraulic oil on the floor can be extremely slippery.
- Remember that when hydraulic fluid is under high pressure and when escaping under high pressure through a small leak, that leak may be invisible.
- Remember that when hydraulic fluid is under high pressure and escaping through a small leak, the escaping high-pressure fluid can penetrate the skin. Injected hydraulic fluid is toxic.
- Remember that hydraulic systems are designed to move large and heavy loads. These loads can be dangerous should they move or shift unexpectedly.
- ALWAYS wear safety glasses.

Key Terms

area	force	Occupational Safety and Health Administration (OSHA)
Blaise Pascal	Joseph Bramah	
Canadian Centre for Occupational Health and Safety	lock out and tag out	pressure
	noncompressible	volume

INTRODUCTION

Modern hydraulic systems perform a wide range of jobs on mobile equipment. From earthmoving to removing trash and refuse to providing services to the disabled, mobile equipment hydraulics play a vital role in a wide range of industries. This book will limit itself to hydraulic system typically found in motor vehicle applications.

SAFETY

Safety Glasses

Most modern repair shops place a great deal of emphasis on eye safety. When hydraulic systems suffer a catastrophic failure such as a ruptured line and fitting, they will usually do so under extreme pressure. These high pressures can turn metal, rubber, and oil into high-velocity projectiles. When working on hydraulic systems, safety glasses are a must **(Figure 1-1)**. If there is concern about the condition of the system, especially pressure-related problems, a ballistic face shield in addition to the safety glasses is strongly recommended.

Skin Penetration Risks

A rather unique, though very real risk when working with hydraulic systems is that of skin penetration. According to the *Occupational Safety Handbook* published by the **Occupational Safety and Health Administration (OSHA)**, pressures as low as 100 psi can force oils and other fluid through the skin **(Figure 1-2)**. In addition to severe lacerations, once these oils enter the bloodstream they often prove to be highly toxic.

Never Open a Pressurized Line

In addition to the potential for eye damage and injection through the skin, a pressurized line may be all that is supporting a major component on a vehicle or mobile hydraulic system that is being repaired. As a personal anecdote, I once arrived to do a class at a city-owned repair shop in California. Only minutes before, a technician had removed a pressurized line from the cylinder on a dump truck bed. The dump truck bed was elevated at the time. The check valve he had assumed would hold the dump bed in the raised position was apparently defective. The dump bed rapidly came down and crushed him. Although he lived, he was severely injured. Never remove a pressurized line or a component from a pressurized line, pump, motor, cylinder, or any other component.

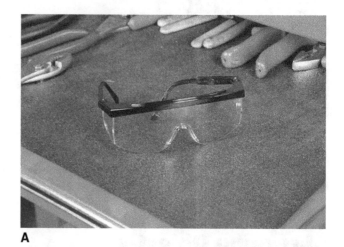

A

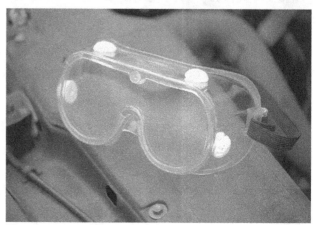

B

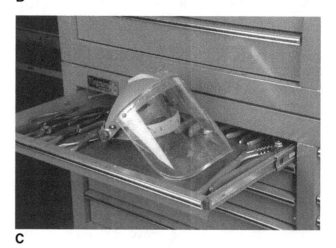

C

Figure 1-1 Everyone gets only a single pair of eyes. Since accidents and unexpected flying parts can occur at any time while servicing fluid power systems, safety glasses should be worn even when performing minor repairs or service.

LOCK OUT/TAG OUT

The purpose of a **lock out/tag out** program is to prevent the accidental operation or start-up of equipment while it is being diagnosed, repaired, or serviced.

Figure 1-2 The Occupational Safety and Health Administration and the **Canadian Centre for Occupational Health and Safety** provide detailed guidelines and procedures to ensure the safety of workers. Procedures and processes are outlined and detailed in various publications of these agencies.

This will protect service technicians and property from damage.

Lock Out/Tag Out Procedures

There are usually formal procedures for lock out/tag out outlined in company employee manuals or company safety manuals. Certain basic procedures always apply **(Figure 1-3)**.

Before working on, repairing, adjusting, or replacing machinery and equipment, the following procedures will be utilized to place the machinery and equipment in a neutral or zero mechanical state.

- Notify all affected employees that the machinery, equipment, or system will be out of service.
- If the machinery, equipment, or system is in operation, shut it down.
- Move switch or panel arms to "Off" or "Open" positions and close all valves or other energy isolating devices so that the energy source(s) (hydraulic pump, etc.) is disconnected or isolated from the machinery or equipment. Accumulators, lines, hoses, and all other components should be relieved of their stored pressure.
- Lock out and tag out all energy devices by using hasps, chains, and valve covers with an assigned individual lock.

Figure 1-3 Locking out the operating controls for fluid power systems ensures that the machinery cannot be operated while maintenance or repairs are being performed. Tagging the lock and/or operating controls provides anyone attempting to operate the equipment with a warning and information about who to contact.

- After ensuring that no employee will be placed in danger, test all lock out and tag out processes by following the normal start-up procedures.
- Machinery or equipment is now locked out and tagged out.
- Should the shift change before the machinery or equipment can be restored to service, the lock out and tag out must remain. If the task is reassigned to the next shift, those employees must perform a review of the lock out/tag out procedure with the previous technician before being allowed to transfer their lock, key, and tag.

CAUTION *After testing, place the controls back in the "neutral" position.*

Restoring Machinery and Equipment to Service

When the task is complete and the machinery, equipment, or process are ready for testing or returned to normal service:

- Check the area to ensure that no employee is exposed to a hazard.
- Account for all tools, repair or replace any defects, and replace all safety guards.
- Remove lock and tag. Restore energy sources. Test to ensure task has been completed satisfactorily.

Procedures Involving More Than One Technician

In the preceding steps, if more than one technician is assigned to a task requiring a lock out and tag out, each must also place his or her own lock and tag on the energy isolating device or devices.

No lock out/tag out program should be built using the steps earlier outlined exclusively. A qualified committee that includes safety professionals should always determine and approve the exact process for any business, school, or agency.

TO GET STARTED

Like many technologies that make significant contributions to our world today, it is impossible to point to one date or person that invented the technology. Hydraulics has much of its beginning in the work of Archimedes, an inventor and scientist from the third century BCE. Some also speculate that the ancient Egyptians may have used hydraulic principles in the building of many of their structures.

Beyond such conjecture, the first patent relating to the practical use of hydraulics was issued to **Joseph Bramah** in 1795 **(Figure 1-4)**. In addition to hydraulics, Bramah was also granted patents on a beverage dispenser, a flush toilet, and a pick-proof lock. All of the technologies in these patents are still in use today. His 1795 patent was for a simple, though effective, hydraulic press. The operation of that press is the principle behind nearly all modern hydraulic systems. In fact, it could be said that today's hydraulic systems are exactly what Bramah patented in 1795, but with additional controls added to make them perform in specific ways.

Bramah's invention of 1795 capitalized on a theory proposed by **Blaise Pascal**, and was in fact a refinement of an invention by Pascal. Blaise Pascal was born in 1623 and actually invented the hydraulic press that was later made practical with the addition of oil soaked leather seals by Bramah a century and a half later. Pascal's most important contribution to hydraulics was the basic operating theory. This theory states that when there is an increase in pressure at any point in a confined fluid, there will be an increase in pressure at all other points in the fluid. His work went on to state that when a force is applied to a piston against the confined fluid, that force will create an increase in pressure within the confined fluid. That pressure change will then act upon another piston with the same change in pressure. The amount of force applied to that second piston by the change in pressure will be proportional to the size of the second piston relative to the first. Double the surface area of the second piston with respect to the first and the amount of force is doubled.

Basic Principles

NATURE OF HYDRAULICS

The term "hydraulics" derives from a Greek word that relates to water. Water and other fluids have several consistent characteristics; foremost among these in the technology of hydraulics is their **noncompressible** nature. It is the nature of liquids to maintain their volume even as pressure is increased upon them. Without this characteristic, a ship sitting on the sea would compress the water below it, the water would give way, and the ship would slip to the bottom of the sea. As it is, the water pushes back as the ship pushes down and the offsetting forces are achieved **(Figure 1-5)**. Modern hydraulics takes advantage of these offsetting forces so that relatively small machines can accomplish relatively large tasks.

The noncompressibility of liquids makes them perfect for the immediate transmission of energy from one place to another. In a closed system, applying a force or setting the fluid in motion produces an immediate and equal response at all other points in the closed system. A related technology, pneumatics, is governed by a very similar set of laws and formulae,

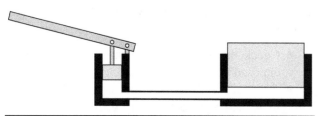

Figure 1-4 In 1795, Joseph Bramah patented a hydraulic press that worked much like the example illustrated here. A small piston, with a small amount of fluid, moved a large piston, multiplying the force many times. His original press was used by the King of England in the Tower of London to flatten paper.

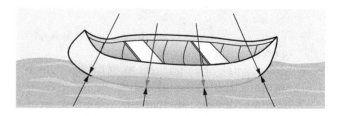

Figure 1-5 The principles of hydraulics have been known for thousands of years. The basic principle can be applied to a boat or ship simply sitting in water, where the weight of the ship applies a force against the water, and the water applies an equal force against the ship. The end result is offsetting forces combine to keep the ship to afloat.

but the "springy" nature of the pressure medium, air, significantly slows the response felt at all other points in a closed system.

SOME DEFINITIONS

Force

Force is the ability to do work or cause a physical change. In a hydraulic system, we think of force as the result of applying a pressure over a given area **(Figure 1-6)**. As a result, the greater the area that a given pressure is applied to, the greater the force. In accordance with that, the greater the pressure applied to a given area, the greater the force. The pound-force, or simply the pound, is actually a measurement of the force required to accelerate a mass of 453.59 grams at a rate of approximately 32.17 feet per second.

In the metric system, the equivalent to the pound is the dyne or the Newton. A dyne is a centimeter-gram-second unit of force. It is equal to the force required to accelerate a mass of one gram at a rate of one centimeter per second.

A Newton is the unit of force required to accelerate a mass of 1 kilogram at a rate of 1 meter per second. A Newton is equivalent to 100,000 dynes.

Area

Area is a measurement that describes the size of a two-dimensional face of a solid object. Area is calculated as a measurement of length, squared. In most

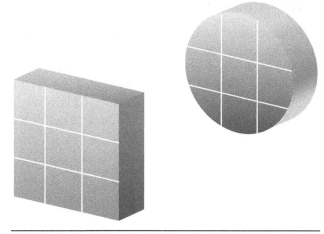

Figure 1-7 Area is a measurement of the exposed face of a surface. In the case of fluid power systems, when a fluid under pressure is in contact with a surface, a force is produced. The amount of the force is equal to the area of the surface times the pressure.

hydraulics used in North America, the measurement used to describe a surface area is square inches. This book will also include metric measurements. The common metric measurement for area is square centimeters **(Figure 1-7)**.

Pressure

Pressure is a measurement of a force applied uniformly over a surface. This measurement is described in units of force per square unit of linear measurement **(Figure 1-8)**. In North American applications, this measurement is usually described in pounds per square inch. The metric measurement is usually stated in *bar*. A bar is a pressure equal to 1 million dynes per square centimeter.

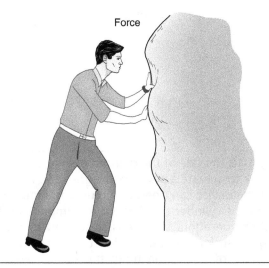

Force

Figure 1-6 Force is the ability to do work. In a hydraulic system, we use force to transfer power from one place to another within the system. Force is a function of pressure being applied over an area.

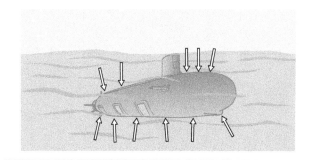

Figure 1-8 Pressure is the force being applied over a unit of area. As a submarine descends into a body of water, the water surrounds the submarine and applies a force equal in all directions. The amount of force acting on a square unit of measure, such as square inches or square centimeters, is called pressure.

Force–Area–Pressure

Perhaps the most critical concept to understand in the design, operation, and troubleshooting of hydraulic systems is the interrelationship between force, area, and pressure. Within Pascal's law there is a set of formulae that describe this relationship. This is often referred to as the FAP, or force (*F*)–area (*A*)–pressure (*P*) relationship.

$$F = A \times P$$
$$A = \frac{F}{P}$$
$$P = \frac{F}{A}$$

Let us say that we apply a pressure of 100 pounds per square over the surface of an object with an area of 100 square inches. The force exerted on that object, and therefore potentially transmitted by that object, is 10,000 pounds of force **(Figure 1-9)**.

Let us take this a step further. Imagine applying these 100 pounds per square inch of pressure over a near cone-shaped object where the surface area of one end is 100 square inches and the surface area of the other is 1 square inch. The force applied to the 100-square-inch surface would be 10,000 pounds. If the 1-square-inch surface was against another object, it would be applying a force equal to the force applied to it on the larger side, 10,000 pounds. But since the force is spread out over an area of only 1 square inch, the pressure applied by the small side to the object it is touching is 10,000 pounds per square inch. Although the pressure is multiplied 100 fold, the force remains the same.

Using this principle, the force of a human hand on the handle of an awl or an ice pick can easily cause the

point to penetrate tough leather. This principle has been applied by humankind since prehistoric times. It is the basic principle behind the stone knife.

Volume–Area–Length

Area also plays a vital role in another principle. Unlike in our cone example, a modern hydraulic system depends on the movement of flow to and through cylinders, and motors through conduits, hoses, or piping **(Figure 1-10)**. All of these devices are three dimensional in nature and therefore not only have area, but contain a volume. **Volume** is the capacity of a region or of a specified container expressed in cubic units. In North American measurement systems, the most common measurement is gallons, the same gallons used to measure milk, water, and other liquids. Cubic linear measurements are often used to describe volume in design and troubleshooting. The most common unit is cubic inches (in^3).

Similarly, the most common metric measurement is liters, although cubic decimeters (dm^3) and other length-related measurements are sometimes used.

Imagine a cylinder with a volume of 2 gallons. Since there is 231 cubic inches in a gallon, we might also say there are 462 cubic inches of volume in the cylinder. If the area of the surface that would be formed across the end of the circle was 12.56 inches, then the length of the cylinder must be 36.78 inches.

A PRACTICAL APPLICATION

In a more practical application, let us calculate the volume of a 6-inch cylinder with a 24-inch stroke.

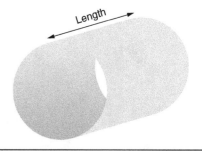

Figure 1-10 The size of the cylinder being used by an object or to actuate a device will impact the speed at which the device can operate. Volume is a measurement of the cylinder that has a direct impact on the speed of operation. Volume is calculated by multiplying the surface area times the stroke or length of movement of the piston times the surface area of the system. If the surface area is 10 square inches and the length of the stroke is 15 inches, then the volume required to stroke the piston from full retraction to full extension is 150 cubic inches.

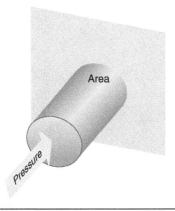

Figure 1-9 The amount of force applied to a surface is equal to the amount of pressure times the amount of area the pressure is being applied over. A pressure of 1,000 psi applied over a surface area of 1,000 square inches equals 1,000,000 pounds of force.

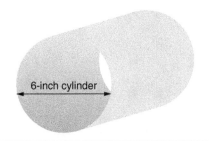

Figure 1-11 When documentation of a fluid power system refers to a 6-inch cylinder, this means that the diameter of the cylinder is 6 inches.

When a cylinder is referred to as a "6-inch" cylinder, it typically means that the diameter of the cylinder is 6 inches **(Figure 1-11)**.

Volume is a function of area multiplied by length. The first thing that needs to be calculated is the cross-sectional area of the cylinder. This is done by multiplying the ratio of a circle's diameter to its circumference, also known as Archimedes constant or pi, by the square of the radius of the cylinder, and then taking that result and multiplying it by the length **(Figure 1-12)**. It sounds more difficult than it actually is. Follow these steps:

1. Measure the diameter of the cylinder. In this case, 6 inches.
2. Divide by 2 to find the radius. This would be 3 inches.
3. Multiply the result by itself. In this case, 3 times 3 equals 9. This process is called *squaring* the number.
4. Multiply the square of the radius, 9, by pi (π). In almost all hydraulic system calculations, pi is rounded to either 3.14 or 3.14159 depending on the level of accuracy required. This would give us a product of 28.26.
5. The cross-sectional area of the cylinder is thus 28.26 square inches.

Once the area is calculated, the volume is found by multiplying the area by the length **(Figure 1-13)**.

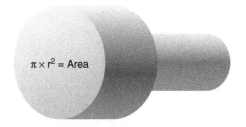

Figure 1-12 The surface area of the piston is equal to half the diameter, also known as the radius, times itself (in other words, squared), times pi (π).

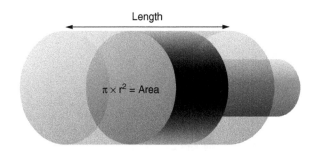

Figure 1-13 Multiplying the surface area of the piston times the length of the stroke will yield the volume of the cylinder.

The result will be in cubic inches. An area of 28.26 square inches multiplied by 24 inches equals a volume of 678.24 cubic inches.

If we want the volume measured in gallons, simply divide 678.24 by the conversion factor of 231. This will yield a calculation of 2.936 gallons. Using these calculations, if we want to move a 6-inch-diameter piston within a cylinder a distance of 24 inches, it would take nearly 3 gallons of fluid moving into the cylinder.

At this point, it is easy to become overwhelmed by the math. Designing a properly operating hydraulic system is a very math-intensive process. Although some math is required at times to get a feel for how a system should be operating, troubleshooting is generally far less dependent on math skills.

Velocity–Flow Rate–Diameter

As important as the amount of force that a hydraulic system can bring to a job is the speed at which that job can be performed. This speed factor is called velocity and it is a direct function of flow rate. Flow rate is a function of the capability of the pump and the diameter of the components, conduits, hoses, and pipes carrying the fluid **(Figure 1-14)**. Flow rate is measured in gallons per minute or liters per minute. Velocity is the time it takes for a mass of fluid or a hydraulic system component to move from point A to point B. Velocity is usually measured in feet per second or meters per second.

Using our previous example of the 6-inch cylinder with a 24-inch length, we will now calculate how long it will take for the piston within the cylinder to move the full range of its stroke. The first thing we will need to know is the supply capacity of the pump supplying fluid to the system. Let us assume that the capacity is 6 gallons per minute. Knowing this flow rate, we can determine how long it should take to move the piston the full length of the cylinder. We have already determined that the fully extended cylinder will hold

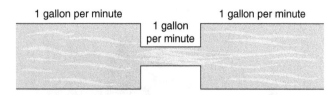

Figure 1-14 The flow rate of fluid in any given branch of a hydraulic circuit is constant throughout the branch. A restriction will cause a change in velocity of the fluid through the restriction. This is similar to a river flowing through a narrow channel. When the channel is wide, the water in the river flows leisurely along. As the channel begins to narrow, the flow rate of the water will remain a constant, but the speed of the water will increase in order to maintain that same flow rate. This is best exemplified by the rapids that are often found at the narrow points of a river channel. Anyone who has ever seen films or videos of white water rafting has seen a perfect example of this.

2.936 gallons. Dividing the extended cylinder volume by the 6-gallon-per-minute flow rate would yield a result of 0.4893 minutes or about 29.36 seconds. Although sometimes manufacturing machinery times must be calculated and monitored that accurately, this book concentrates on mobile equipment hydraulics and therefore it is acceptable to round the 0.4893 minutes to half a minute or 30 seconds.

The preceding calculations assume that whatever the pump produces can arrive at the cylinder at the same rate the pump is sending it. As we will see in subsequent chapters, this is seldom the case. The diameter of the hoses leading to and from can limit the flow rate to the cylinder. In general, the smaller the diameter, the lower the flow rate.

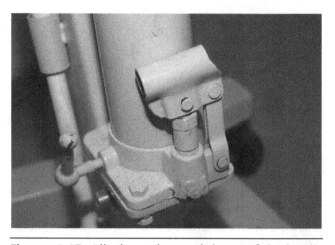

Figure 1-15 All the rules and laws of hydraulic systems and fluid power systems are at play even in this simple bottle jack. The user applies a small amount of work many, many times to the small piston on the side of the jack. Each time the work is applied, a small amount of fluid is moved from the small system to the large piston. Slowly, the extend side of the large piston is moved upward by each sequential movement of the small piston. The pressure applied by the small piston to the large piston is multiplied in force by the surface area of the large piston, allowing the bottle jack to move or lift very heavy loads.

Decreasing the diameter causes a decrease in flow rate, and as it causes this decrease in flow rate, another characteristic of the fluid flow called velocity will increase. As the diameter of the hose decreases, the speed of a given drop of fluid must increase in order for the flow rate to be maintained **(Figure 1-15)**. This increase in speed is an increase in velocity. There will be more on that in later chapters.

Summary

Hydraulics is both an ancient science and a new technology. Like any technology, many complexities and nuances can affect the operation of a hydraulic system. In spite of that fact, all of the complexities and nuances follow a rather small number of laws and rules and are governed by a very small set of mathematical principles. Understanding these principles, regardless of one's skill in mathematics, will make a technician efficient and skilled when troubleshooting and repairing hydraulic systems.

Review Questions

1. Applying an even pressure on the large end of a cone will result in what kind of pressure at the small end of the cone?

    A. A much greater pressure   C. The pressure will not change

    B. A much smaller pressure   D. Only the force will change

2. When fluid flows into a cylinder, as the flow rate increases, the speed of the piston in the cylinder:

    A. Increases.

    B. Decreases.

    C. Is unaffected.

    D. Whether it increases or decreases depends on the diameter of the cylinder.

3. A pressure of 200 psi is applied to one end of a piston that has a surface area of 10 square inches. What is the total force on the piston?

    A. 20 pounds of force

    B. 200 pounds of force

    C. 2,000 pounds of force

    D. This cannot be calculated with the information provided.

4. In the preceding question, if the other end of the piston has a surface area of 100 square inches, what would be the pressure on this larger end?

    A. 2,000 psi

    B. 200 psi

    C. 20 psi

    D. 2 psi

5. If all other factors remain the same, increasing the surface area of a piston:

    A. Tends to increase force.

    B. Tends to decrease force.

    C. Tends to decrease pressure (psi).

    D. Both A and C

6. In the United States, the agency responsible for establishing guidelines related to employee safety is:

    A. United States Safety Department.

    B. Office of Management and Safety.

    C. Occupational Safety and Health Administration.

    D. No such agency exists.

7. When a submarine submerges, the water applies a pressure around the cylinder of the submarine that is:

    A. Equal at all points around the cylinder.

    B. Greater on the top than it is the bottom.

    C. No pressure is applied; there is only force.

    D. The force is equal, but the pressure changes along the length of the cylinder.

8. The lock out/tag out procedure is important:

    A. To protect the technician from personal injury.

    B. To protect the equipment from damage.

    C. Both of the above

    D. Neither of the above

9. Personal protective equipment (PPE) is extremely important in providing a safe working environment for the technician. Who is responsible for ensuring that every technician properly wears protective equipment?

    A. The individual technician

    B. The technician's employer

    C. The technician's co-workers

    D. All of the above

10. A simple hydraulic bottle jack follows all the rules of fluid power and hydraulic systems except for:

    A. Flow, area, pressure.

    B. Area, pressure, velocity.

    C. Flow, velocity, area.

    D. It follows all the rules of fluid power and hydraulic systems.

CHAPTER

2

Building a Basic Cylinder or Hydraulic Motor Circuit: A Non-Math Explanation of Operation

Learning Objectives

Upon completion and review of this chapter, the student should be able to:

- Describe fluid flow through a typical cylinder-based hydraulic system.

- Describe fluid flow through a typical motor-based hydraulic system.

- Predict how pressure changes as fluid flows through a typical hydraulic system.

- Predict how temperature changes as fluid flows through a typical hydraulic system.

- Predict which components in a typical hydraulic system can affect flow.

- Predict where and how velocity will change in a typical hydraulic system.

- Describe how changes in pressure cause changes in force.

- Describe how changes in flow rate cause changes in the speed of operation in a system.

Cautions for This Chapter

- When inspecting a system, remember that the hydraulic fluid may be under high pressure.

- When working around a system with leaks, remember that hydraulic oil on the floor can be extremely slippery.

- Remember that when hydraulic fluid is under high pressure and when escaping under high pressure through a small leak, that leak may be invisible.

- Remember that when hydraulic fluid is under high pressure and escaping through a small leak, the escaping high pressure fluid can penetrate the skin. Injected hydraulic fluid is toxic.

- Remember that hydraulic systems are designed to move large and heavy loads. These loads can be dangerous should they move or shift unexpectedly.

- ALWAYS wear safety glasses.

Key Terms

directional control valve   hydraulic motor   restriction to flow

hydraulic cylinder   pressure relief valve

INTRODUCTION

In this chapter, we will be using terms and discussing the components commonly found in a simple hydraulic system. To accomplish that, we will take a journey through this simple hydraulic system, traveling from component to component and discussing how each component affects the performance of the system and how the components interact **(Figure 2-1)**. On our journey, we will see how the fluid flows, where it does work, where its temperature changes, where it can escape from the system (leak), what impedes the flow, what changes the fluid's velocity, what damages the fluid, and what contaminates the fluid.

In the mobile equipment industry, the terms "hydraulics system" and "fluid power system" are often used interchangeably. Of the two terms, the term "fluid power system" most accurately describes what the motor vehicle service technician will find installed on commercial vehicles, construction equipment, and vocational trucks. Fluid power systems are installed on these vehicles to ease the workload and reduce the manpower required to complete a task. In many cases, the truck itself is little more than a transportation

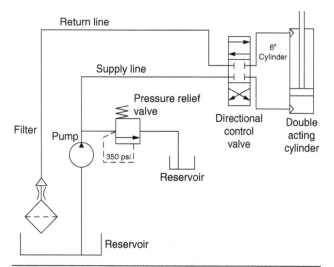

Figure 2-1 Modern hydraulic systems range from the very simple to the very complex. In this relatively simple system, flow from a pump is routed through a directional control valve to a double acting cylinder. When the directional control valve is in one position, the cylinder extends; when the valve is moved to the opposite position, the cylinder's piston retracts.

device for the fluid power system. An example of this would be what is often referred to as a bucket truck or cherry picker. These are more formally called "truck-mounted aerial lifts." The purpose of the vehicle is to get a worker 20, 30, or maybe 150 feet into the air in a safe work platform. The purpose of the truck is just to provide a convenient method of transport for the aerial platform, as well as a base for that aerial platform. In some cases, the truck or construction vehicle is actually a part of the tool. This is the case with a dozer, or a snowplow. The combination of the vehicle and the fluid power system can move hundreds of cubic feet of snow in a matter of minutes or even seconds; a volume of snow that would take a crew of hundreds of workers hours or even days to accomplish. In fact, there are parts of the world where roads would be completely impassable in the winter because no human crew could work as fast as the snow would fall.

Fluid power, or hydraulic systems, utilize one of the fundamental precepts of physics called the law of conservation of energy. This law states that the total amount of energy in a closed system remains constant. The implication is that energy could be put into a system and the system could perform with that "trapped" energy indefinitely. In effect, this would form a perpetual motion machine. What prevents this from being possible is that as any machine operates it is affected by its own internal resistances. The internal forces that affect the inability of a fluid power system to be a perpetual motion system fall into two categories: resistance to flow and heat. Although the fluid usually has excellent locating qualities, it still requires energy to push the fluid through the system. It requires energy when there is a resistance to flow. If there is a resistance to flow, energy will be consumed and some of that energy will be converted into heat. Environmental factors such as dust and humidity can also add to the amount of friction, resistance to flow, and heat with which the system has to contend.

Short of being perpetual motion machines, fluid power systems use a power source to operate a pump with the intent of conserving as much of the energy from the power source as possible to perform a task in a location remote from the power source. For instance, in a typical mobile fluid power system, the vehicle's engine is the power source and conservation of energy allows the hydraulic fluid to transfer that energy with

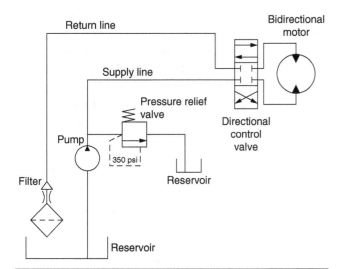

Figure 2-2 Although the hydraulic cylinder is the most common device to be powered by a fluid power system, another popular device is a hydraulic motor. In this example, a bidirectional hydraulic motor is used. When fluid is routed through the directional control valve's upper passages, the motor will rotate in one direction; when the valve is moved to the opposite position, the motor will rotate in the opposite direction.

minimal loss to a machine performing work at a different location on a vehicle.

In the circuit that is built in this chapter, the pump will be used to transfer energy to a **hydraulic motor** **(Figure 2-2)**. Along the way, work will be done and energy will be lost in the form of heat. The movement of fluid through a system to perform work is far more conserving of energy and far more efficient than systems that use gears, chains, cable, pulleys, drive shafts, and various connecting joints to perform the same tasks. All of the strictly mechanical methods of transferring force and energy from one place to another are rotating devices that waste energy through vibration, and cable-like devices that waste energy through distortion and stretching. These devices are also prone to wear and require a great deal of maintenance.

Fluid power systems require far less maintenance than their mechanical counterparts. In part, the reason for this is related to less energy being lost during operation to vibration and heat. Fluid power systems are simpler, less wasteful, and far more durable than their mechanical predecessors. Soon after the turn of the last century, steam shovels, earthmovers, and even factory-mechanized systems depended on cables, belts, and drive shafts to transfer energy. In the latter half of the last century, these devices were replaced by fluid power systems as a lighter weight and more dependable alternative.

RESERVOIR

Purpose of the Reservoir

The reservoir serves several roles in the typical hydraulic system. The first and perhaps most significant job of the reservoir is to store the hydraulic oil. The reservoir will usually hold many times the amount of fluid that can be in the rest of the hydraulic system at any given time **(Figure 2-3)**.

The reason there is so much fluid in the reservoir is to allow the fluid an opportunity to rest. Resting the fluid might seem like an unusual concept for an inanimate substance such as hydraulic oil but it really is not. The fluid returning from its journey through the working part of the hydraulic system is warm, probably has had some air churned into it, and may have some contaminates. When this oil returns to the reservoir, its velocity slows to virtually zero. The fluid is given a chance to remain stationary. Heat can be dissipated into the air covering the top of the fluid and to the walls of the reservoir **(Figure 2-4)**. In much the same way as a fast food restaurant milk shake will separate when left on the kitchen counter overnight, aerated oil will give up the trapped air as it lays relatively motionless in the reservoir.

While in the reservoir, the relatively motionless hydraulic oil also has an opportunity to allow any contaminants that may have made it through the filter or that may have bypassed the filter to settle out of the fluid. While this is a very inefficient filtration method,

Figure 2-3 Hydraulic reservoirs come in an almost infinite number of shapes, sizes, and materials. A common material used in mobile hydraulic systems is steel. It is relatively lightweight, very strong, and does a good job of transferring the heat of the fluid returning from the working system to the surrounding air. Like many fluid power systems found on mobile equipment, this reservoir features a sight glass for easy observation of the fluid level.

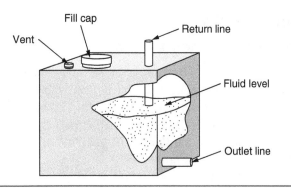

Figure 2-4 When the fluid returns from the working part of the system, it has usually gained heat. The reservoir is usually designed with enough capacity to allow the returning fluid to greatly decrease its velocity and thus rest. This provides a time for the fluid to dissipate some of its heat into the surrounding air. The relatively static state of the fluid also provides an opportunity for foreign materials, particulate matter, and hard particle contamination to settle out of the fluid and not pass through the system.

it does offer some protection against contamination. The larger the size of the reservoir and the larger the amount of hydraulic oil in the reservoir, the slower the fluid moves through the reservoir and the greater the system's ability to separate contaminants. In all cases, except cases of the poorest maintenance, the primary entry point for contamination is through the filler opening in the top of the reservoir. Every time the filler is opened, the opportunity arises for sand, dust, metal shavings, and a wide range of other damaging contaminants to enter the hydraulic oil. Since the filtration system is almost always on the return to the reservoir, these contaminants get to make at least one trip through the hydraulic system. This settling process offers a small amount of protection from lack of care in cleaning the area around the filler when the system's reservoir oil level is being serviced.

FLUID FLOW IN THE RESERVOIR

Hydraulic oil (also called hydraulic fluid) can enter the reservoir in two places. The first is the already mentioned filler port. Usually a screw-on cap covers a large opening to service the oil reservoir. This is where the oil is added by the vehicle operator or service technician. This is also the first and best opportunity for contamination to enter the system.

The other place the oil enters the reservoir is through the return line. A filter usually is put in place on the way to the reservoir. This filter removes any contaminants that may have made it this far through the system. Once the returning oil arrives in the reservoir, its velocity

drops to nearly zero. This extremely low velocity permits contaminants to settle to the bottom of the reservoir, and trapped air to escape from the returning oil.

The reservoir usually has two exit lines. The first is a drain port used to remove oil from the reservoir during a routine change of the hydraulic fluid. In some cases, this line is also used to purge water, metal chips, and other heavy contaminants from the system.

The second, and actually the primary, exit line is the supply line to the pump. This exit carries the full supply of hydraulic fluid required to make the machine operate. If the amount of oil that leaves the reservoir is greater than the amount that returns, the system has a leak. Generally speaking, the outlet port on the reservoir will be higher than the inlet to the pump. This will allow gravity to assist atmospheric pressure to move the oil to the pump.

HOW THE RESERVOIR PERFORMS WORK

Above the reservoir is a column of air about 100 miles high **(Figure 2-5)**. This column supplies a force of about 15 pounds per square inch to the surface of the hydraulic oil sitting in the reservoir. This means there will be a predictable pressure at the outlet of the reservoir **(Figure 2-6)**. This is because whatever pressure is exerted on one plane of the reservoir, the top surface of the oil will be the same pressure as that

100-mile column of air creates a pressure of 14.7 pounds on each square inch of the surface of the oil in the reservoir.

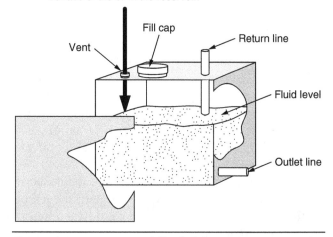

Figure 2-5 In many mobile fluid power systems, the only force acting on the hydraulic oil to move the oil to the pump is the force of atmospheric pressure. The vent in most mobile fluid power systems is critical. It is more than a matter of "the fluid cannot flow out if the air cannot flow in." It is through the vent that the force of atmospheric pressure is applied to the hydraulic oil. This is the force that moves the oil toward the pump.

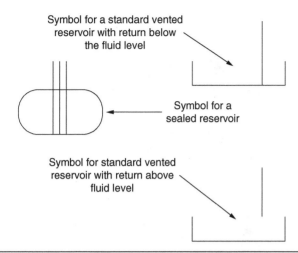

Symbol for a standard vented reservoir with return below the fluid level

Symbol for a sealed reservoir

Symbol for standard vented reservoir with return above fluid level

Figure 2-6 These symbols represent common varia-tions of reservoirs used in mobile fluid power. The one on the upper right is the most common. It is vented to the atmosphere to allow atmospheric pressure to push the hydraulic oil to the pump. The reservoir on the left middle is a pressurized reservoir. Pressurized reservoirs are used when the reservoir is located far below the pump inlet or when the vehicle is likely to be partially submerged. Since pressurized reservoirs are sealed from the atmo-sphere, they are also sealed from water and other contamination.

exerted on all other planes. This includes the plane on which the outlet to the pump is located.

In addition to the pressure exerted on the surface of the hydraulic oil in the reservoir, there is the weight of the oil itself. If you choose a 1-square-inch column of oil and measure a foot of that column, it will weigh approximately 0.4 pounds, or about 6.4 ounces. Therefore, if the fluid level in the reservoir is 2 feet deep, the fluid will add about 0.8 pounds per square inch additional pressure, bringing the total pressure to about 14.8 pounds per square inch.

Since the inlet of the pump also has about 14 pounds per square inch applied to it, the additional pressure provided by the depth of the fluid in the reservoir plays an essential role in moving the fluid toward the inlet side of the pump. Low fluid level can prevent proper movement of the hydraulic oil toward the pump, this can cause the pump to be starved for oil, which can cause it to be starved for lubrication and lead to the premature failure of the pump.

TEMPERATURE CHANGES IN THE RESERVOIR

When the oil returns from the system, it is always warmer than when it left the reservoir. Its journey through the hydraulic system causes it to absorb heat from operating components, from friction as the fluid travels through hoses, and from the changes in

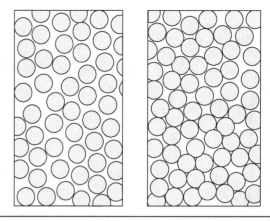

Figure 2-7 Heat is generated in hydraulic fluid when pressure attempts to force molecules of the fluid together. These molecules begin to rub on one another, generating friction and therefore heat.

pressure that the fluid experienced during its journey. When the fluid enters the reservoir and has an op-portunity to rest, it is able to give up some of its heat. The larger the reservoir and the greater the amount of fluid in the reservoir, the greater the amount of heat the hydraulic fluid is able to give up **(Figure 2-7)**.

Hydraulic oil, like most oils, is slow to absorb and give up heat. The tendency of the oil to be slow in absorbing heat is one of the significant reasons it is used as a pressure medium in hydraulic systems. The down side of that advantage is that it is also slow to radiate, or give up, heat. Typically, hydraulic oil stored in a steel reservoir has a heat dissipation factor of 0.001 BTU per square foot of surface area for every degree of temperature differential between the oil and the air per hour.

Let us assume the temperature of the oil in the reservoir is 30° above the air temperature. Let us also assume that the surface area of our steel reservoir is 27 square feet and that the reservoir is 80 percent full. That means the hydraulic oil is in contact with 21.6 square feet of the reservoir's surface area.

The area of 21.6 square feet, times the temperature difference of 30°F, times the dissipation factor of 0.001 yields a dissipation rate of 0.648 HP.

Using the conversion factor of 33,445.7 BTUs per hour equals one horsepower, 21,673 BTUs per hour would be dissipated. A result like 21,673 BTUs is hard to grasp as a concept.

This is equivalent to the amount of heat energy in about one fifth of a gallon of gasoline.

The important part of all this is that as the tem-perature differential decreases, the cooling capacity of the reservoir also decreases.

Also, if the reservoir is only filled to half its normal full level, the hydraulic oil would only be touching

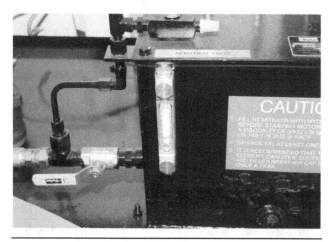

Figure 2-8 It is important for the vehicle operator to check the fluid level in the fluid power system of his vehicle on a regular basis. Low fluid levels can lead to excess heat in the system, pump cavitation, foaming (aeration) of the hydraulic oil, and poor lubrication of the moving parts of the system.

Figure 2-9 A damaged reservoir can leak hydraulic fluid. In addition to creating a mess, hydraulic oil may not be compatible with materials used in other systems on the vehicle. Wheels, tires, drive belts, wiring insulation, brake linings, and even battery cases can be damaged by some hydraulic fluids.

about 40 percent of the surface area and the cooling capacity will be cut by 50 percent **(Figure 2-8)**.

LEAKAGES IN THE RESERVOIR

Although many hydraulic systems use pressurized reservoirs, in mobile equipment they are the exception rather than the rule and therefore will be discussed in a subsequent chapter. Reservoirs are typically made of steel, high-impact plastic, or composite materials. Leaks in the reservoir are usually quite evident **(Figure 2-9)**. After the vehicle sits, there is an oily patch below the reservoir on the ground. Another significant indication of a leak in the reservoir is a film of oil or dirt-laden oil streaming from the site of the leak.

FLOW IMPEDANCE IN THE RESERVOIR

Other than damage to the reservoir return port, there is very little about the reservoir itself that can reduce or impede the flow of fluid into the reservoir. Since most mobile equipment uses a reservoir that is open to the atmosphere through a vent or vent valve, changes in fluid level in the reservoir cause air to be moved in and out of the reservoir. When the level of the fluid rises, air needs to be pushed out of the reservoir to make room for the fluid or the tank will become pressurized. If the vent is restricted, as the fluid returns to the reservoir it tends to pressurize the trapped air. This pressure impedes the flow of returning oil and decreases the efficiency of the entire system.

When the system withdraws oil from the reservoir, the fluid level tends to decrease. The reservoir depends on the same vent that allows air to be pushed out of the

tank to also allow air to be drawn into the tank to replace the departing fluid. If the vent is restricted, this can keep the outside air from entering **(Figure 2-10)**. As the fluid level drops, the pressure of the air on the

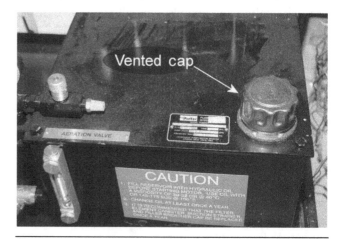

Figure 2-10 Although many reservoirs will have separate vents, many have the vent in the cap. Since this is where air enters the reservoir to push the fluid toward the pump, this vent must be filtered. In this system, the vent is in the reservoir fill cap. These caps should be routinely inspected to ensure that the filtration material is in place and is not damaged or contaminated.

surface of the hydraulic oil decreases, which may cause pump cavitation, or an insufficient supply of oil to the pump. This can damage the pump.

Theoretically, the amount of fluid being withdrawn from the reservoir is exactly the same as the amount of fluid returning to the reservoir. This is true assuming there are no leaks in the system, and assuming we are referring to the mass of the fluid. The increases in temperature of the returning fluid, however, cause its volume to be slightly larger. This means that air will need to be vented to the atmosphere. More critical than this is that the returning warm oil will heat the air in the reservoir causing significant expansion of the air. The expanded air must be vented or the reservoir will pressurize. Similarly, when the system is shut down, the fluid, and especially the air, will cool and contract. If the vent is restricted, this can cause a partial vacuum in the reservoir, making it very difficult to supply the pump with the required supply of oil upon starting the system.

FLUID VELOCITY CHANGES IN THE RESERVOIR

It is important to remember that the amount of flow will be consistent at all points in the hydraulic system **(Figure 2-11)**. The only exception is the points in the system where the fluid divides to follow parallel paths. However, if the amount of fluid in the parallel paths was measured, it would equal the amount of fluid that would be flowing in any other non-parallel portion of the hydraulic circuit. The reservoir is no exception to this rule: whatever flows out will flow back in. The volume of hydraulic oil in the reservoir should be large to allow for proper cooling and sedimentation.

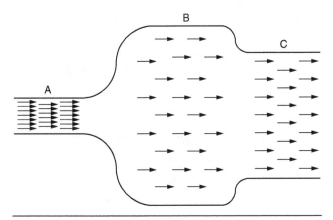

Figure 2-11 In any hose, line, or section of the system, the flow rate (gallons per hour or gallons per minute) is the same. If the hose or other passageway narrows, the velocity of the hydraulic oil must increase in order to maintain the same flow rate. If the hose or passageway widens, the velocity of the fluid at a given flow rate can, and in fact must, decrease.

The flow rate in all parts of a hydraulic circuit is the same as the flow rate in all other parts of the circuit. This includes the flow rate through the reservoir. In the Canadian Province of British Columbia flows a large river called the Fraser River. From its source in Alberta, it flows hundreds of miles before it reaches the Pacific Ocean. As it travels, tributaries, rivulets, creeks, and runoff add to the flow of the river. By the time it passes through the city of Prince George it has completed less than half of its journey to the sea. At this point, the Fraser is already several hundred feet wide. The river is rolling along at a rather leisurely velocity. By the time the river is a couple of hundred miles south of Prince George, the volume of water during peak flow in the spring is over 200 million gallons of water per minute. It is here that this wide leisurely river passes through a gap between two mountains. Here all of the leisurely traveling water, all 200 million gallons, must pass through a 110 foot wide gap known as Hell's Gate. The only way this flow rate can be maintained is by increasing the velocity of the flow. Once the river is downstream of Hell's Gate, it returns to its leisurely flow.

At a given flow rate, the smaller the passageway for the hydraulic oil, such as a small hose, the greater the velocity or speed of the oil. The larger the passageway for the oil, such as the reservoir, the slower the velocity of the oil. The slow velocity of the oil in the reservoir allows for greater cooling of the oil while it is in the reservoir. The low velocity allows contaminates and heavy particles that might have made it through or bypassed the filter to settle out.

FLUID DAMAGE IN THE RESERVOIR

The reservoir is mostly a safe haven for the hydraulic oil. No damage can occur here. It is allowed to cool and to settle out its contaminants.

FLUID CONTAMINATION IN THE RESERVOIR

Although a safe haven from damage, the reservoir is not a completely safe haven from contamination. When the operating or servicing personnel remove the filler cap, any dirt or contamination on or around the filler cap has an opportunity to fall into the reservoir. Always make sure the area around the cap is clean and free of debris before removing the cap (see **RS 1-1, p. 40**).

THE SUPPLY LINE
Purpose of the Supply Line

The supply line delivers the hydraulic fluid to the pump. In most systems, the only pressure in this line is the force of atmospheric pressure and gravity acting on

Figure 2-12 Since the pressure and heat stresses on the suction line are minimal, they can be made of lower-quality materials than many of the other hoses in the fluid power system. In spite of this, the materials that are usually used are of relatively high-quality. As the primary supply line of the system, a hole or damage to the hose can cause air to be pulled in, resulting in poor fluid flow and possibly pump cavitation.

the fluid in the reservoir. Since the pressure at this point is generally very low, these hoses can be made of materials that could not handle higher pressures **(Figure 2-12)**. The supply line may be made of rubber, metal, or even plastic.

FLUID FLOW IN THE SUPPLY LINE

It is commonly believed that the pump pulls the fluid through the supply line to begin the fluid's journey through the system. The inlet side of the hydraulic pump does create a low-pressure area as the pump operates, but this low-pressure area is so weak it cannot possibly draw the hydraulic oil into the pump on its own **(Figure 2-13)**. Fortunately, a 100-mile-tall column of air is applying a force to the surface of the hydraulic oil. The pressure acting on the surface of the oil is a little under 15 psi, and less at higher altitudes. If the surface area of the oil in the reservoir is 10 inches by 40 inches, the force pushing the fluid from the reservoir toward the pump is 6,000 pounds. This force is equally and evenly applied across the surface of the oil in the reservoir. It is also acting vertically on the fluid. This force equates to the surface area of the hydraulic oil times the force of the air column above the reservoir. The oil then presses out equally against all surfaces of the reservoir. If the reservoir is 10 inches by 40 inches and the oil in the reservoir is 12 inches deep, then the surface area that the force of the oil is acting upon is equal to 10 inches by 12 inches times 2. This is an area of 240 square

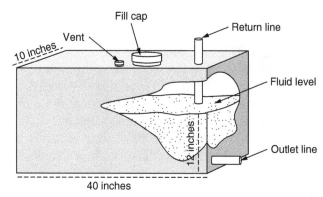

Figure 2-13 The pressure of the fluid at the reservoir outlet is determined to a large extent by the dimensions of the reservoir. A force applied to the surface of the fluid in the reservoir is transferred to and spread out across the surface of the bottom of the reservoir and the sides of the reservoir in contact with the fluid. Since the fluid's contact area with the bottom and walls of the reservoir is greater than the surface area of the fluid in contact with the atmospheric pressure, the pressure on the walls of the reservoir and the pressure at the outlet port of the reservoir will always be much lower than the atmospheric pressure.

inches and accounts for the area of the two narrow walls of the reservoir that the hydraulic oil contacts. The area of the wide walls equates to 40 inches times 12 inches times 2. This is 960 square inches. The bottom of the reservoir is 40 inches by 10 inches or 400 square inches. The total surface area that the oil is contacting is 1,600 square inches. When the 6,000 pounds of force acting on the surface area of the oil is spread across the 1,600 square inches of the reservoir in contact with the oil, it equates to pressure acting on the walls and bottom of the reservoir equal to 3.75 pounds per square inch (psi). This then acts as the pressure to move the fluid toward the reservoir. Couple this with the slight suction provided by the inlet side of the pump and fluid will begin to flow.

HOW THE SUPPLY LINE PERFORMS WORK

The supply line is simply a passageway for fluid to pass from the reservoir to the inlet of the pump. Forces applied to the fluid in the reservoir supply the energy that forces the hydraulic fluid through the supply line. The fluid is therefore pushed through the hose, not pulled through the hose by the pump.

TEMPERATURE CHANGES IN THE SUPPLY LINE

Since only minimal forces are applied to the fluid as it passes through the supply hose, there will be little or no increase in temperature. Also, since the surface area of the supply hose is relatively small compared to the volume of oil passing through the hose, very little cooling of the fluid will occur.

LEAKAGES IN THE SUPPLY LINE

One of the prime places for a hydraulic system to develop a leak is in its hoses. Many of the hoses carry very high pressures and are expected to remain intact and leak free at those pressures. The supply hose in most systems is subjected to very little pressure or heat. In many cases, the supply line is subject to vibration, however. Most deterioration and leaks come from the supply line not being properly secured or equipped with proper strain relief **(Figure 2-14)**.

FLOW IMPEDANCE IN THE SUPPLY LINE

The most common defect or fault in the supply line that impedes flow is a kink in the line. Other conditions that can result in supply line restriction are internal deterioration of the hose and foreign object blockage.

FLUID VELOCITY CHANGES IN THE SUPPLY LINE

In almost all cases, the supply line has a constant inside diameter. There are no restrictions or enlargements

Figure 2-14 This hose shows evidence of deterioration on its outer protective layer. Although this damage is not evidence of damage to the interior of the hose, when coupled with evidence on leakage it could indicate that the hose is in need of replacement. The heat in the lines to and from the actuator device carries a lot of pressure. This can cause minor seepage through the walls of the hose, which does not require immediate replacement of the hose. However, eventually the heat will deteriorate the hose from the inside out. Leakage at a metal fitting may indicate deterioration of the interior of the hose and will therefore warrant replacement.

for the fluid to pass through on its way to the pump. As a result, the speed of the fluid does not change as it passes through the line.

FLUID DAMAGE IN THE SUPPLY LINE

A constant velocity and a constant flow rate at low pressure mean that there is a constant temperature in the supply line. Assuming proper fluid levels in the reservoir and proper system design, the fluid is relatively cool, thus the fluid itself will cause no damage to the supply line. Additionally, if the proper material is used for the supply line, that is to say if it has not been replaced with the wrong material, the relatively cool supply fluid should not degrade the hose material and therefore the fluid should not be damaged by degraded hose material.

FLUID CONTAMINATION IN THE SUPPLY LINE

There is little opportunity for the fluid to become contaminated in the supply hose. The only significant possibility is that the wrong supply hose material may have been used. As this improper material ages, coupled with exposure to the hydraulic oil, it may begin to shed. Small particles of the inner hose may begin to shed or flake off, thereby contaminating the fluid on its way to the pump.

THE PUMP

Purpose of the Pump

The pump might be referred to both figuratively and literally as the heart of the system. The supply line delivers the hydraulic oil to the inlet of the pump. The gears or pistons of the pump draw the fluid from the inlet into the pumping chambers. This process and its many variations will be covered in the next chapter.

Although the pump does create a slight suction on the supply line, its primary task is to push fluid through the system **(Figure 2-15)**. When it comes to troubleshooting, it is important to remember that the pressure created by the pump, regardless of the type of pump or the type of hydraulic system, is zero pounds per square inch. Pumps create volume, they create flow, and they do not create pressure. When the flow generated by the pump encounters a restriction, then the push of the fluid against that restriction causes the pressure of the fluid to increase. If the restriction is reduced, then the pressure decreases **(Figure 2-16)**.

In troubleshooting a hydraulic problem, it is important to remember the pump is rarely the cause of low pressure in the system. When the pump is sufficiently defective to cause low pressure in the system, it

Figure 2-15 The pump is the component in a fluid power system that creates flow. Although many times a technician will look to the pump as the source of a pressure-related problem, most of the time it is not the location of such problems. Flow capacity of the pump is determined not only by the size of the pump, but also by the speed at which the pump is being turned.

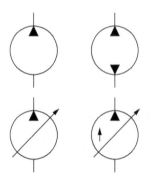

Figure 2-16 These symbols are used in fluid power diagrams to represent pumps. The symbol on the upper left is a fixed displacement pump such as the one being used in the example. The upper right is a bidirectional pump capable of moving fluid in either direction. The symbol on the lower left represents a variable displacement pump. These pumps are capable of varying their flow rate even when the rotational speed of the pump remains constant. The symbol in the lower right corner represents a variable displacement pressure compensated pump. These pumps alter their flow rate based on changes in the pressure within the system.

has long since been incapable of providing sufficient flow for the system to function at a proper speed.

Although it may be a bit simplistic to say, pressure in the system determines the power in the system. Pressure is directly related to the amount of work the system can do. Flow is related to how fast the system

operates. The capacity of the pump has a direct impact on flow and therefore a direct impact on how fast the system can operate.

FLUID FLOW IN THE PUMP

As previously mentioned, the pump is the heart of the hydraulic system. Fluid is drawn in from the low-pressure area as the inlet is moved into the gear chambers, vane chambers, or piston cylinders. Rotating gears or ascending pistons then accelerate the fluid toward the outlet of the pump.

The efficiency of the pump can be measured in two ways: the volumetric efficiency and the thermal efficiency. The volumetric efficiency relates to the ability of the pump to fully fill its gear chamber, vane chambers, or cylinders and then push all of the hydraulic oil that filled these areas into the system. When the pump is new, the volumetric efficiency is very high, usually approaching 100 percent. As the pump is used, wear begins to occur and it may not be able to fully fill the inlet chamber or cylinders. As a result, the amount that the pump actually draws in on each cycle is less than what a new pump would be able to draw into the inlet. This means that the volumetric efficiency is decreased (**Figure 2-17** through **Figure 2-21**).

Another cause of decreased volumetric efficiency is a decrease in the pump's ability to push fluid against the restrictions to flow in the system. This results from internal bypassing caused by wear inside of the pump. As the pump gears, pistons, or vanes attempt to push

Figure 2-17 A basic gear-type hydraulic pump consists of two meshing, counter-rotating gears. A slight suction on the inlet side draws the fluid into the pump. The rotating gears pick up the hydraulic oil and accelerate it through the meshing of the gears and into the outlet chamber. The faster the pump is rotated, the faster it moves the fluid and therefore the greater the flow rate of the pump.

Figure 2-18 The gears of the pump used in the example system are closely toleranced and precision-machined. The flow rate is completely dependent on the size of the pump and the speed at which it is being rotated. These pumps are extremely simple and therefore very dependable.

Figure 2-19 This is a variation of the gear pump. Called an internal gear pump, they offer quieter operation. As the inner and outer gear mesh, the hydraulic oil is squeezed and accelerated forward through the system.

Figure 2-20 This is a cutaway of a vane type pump. Sections of the pump have been removed for instructional purposes. It consists of a hub riding on a large shaft. The hub has grooves cut into the outer machined surface. Vanes, or scrapers, ride in these grooves and press against the machined walls of the chamber in which the hub rides as the hub is rotated. The photo shows a chamber that has been cut away on two sides to make the vane grooves more visible. These pumps offer quiet operation with low power consumption.

Figure 2-21 The variable displacement piston pump uses an array of pistons mounted in a set of rotating cylinders. The end of these pistons ride on a device called a swash plate. Even when the speed of the pump is held steady the volumetric flow can be altered by changing the angle on the swash plate. The swash plate angle can be altered in many ways. In the case of this pump, the angle is varied by means of the lever at the top of the photo.

the fluid against the resistance, the fluid finds that it is easier to slip past the gear, piston, or vane internally than it is to move against the **restriction to flow** in the system. This creates a low flow volume and with extended wear can even cause low pressure in the system.

When the fluid in the pump bypasses the gears, pistons, or vanes, then heat is generated. This heat is a result of the friction of the fluid as it slides through the worn areas. This heat in turn can damage the fluid and components inside the pump.

Thermal efficiency is as important, but more difficult to describe. All work is a function of the utilization of energy. Energy, when it does work, generates heat.

When a piece of wood is burned, energy is released as heat. When hydraulic fluid moves though a system, it does so to perform work such as to lift or to move. On its journey through the system it encounters components and objects that impede its flow. These encounters

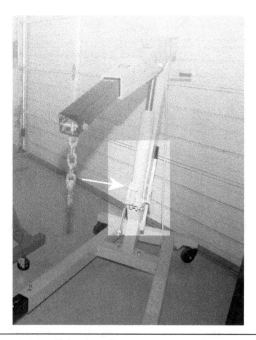

Figure 2-22 All hydraulic pumps must have a power source. In mobile equipment, the most common sources are electric motors, gasoline engines, and diesel engines. This engine crane is an example of a very simple hydraulic system. The manually operated pump is shown in the highlighted area.

generate heat because of friction. This heat does not come without a price. Any heat that is generated is energy that is not used to lift or move. Thermal efficiency is the percentage of energy used to do the moving or lifting and is not converted into heat. A pump with a thermal efficiency of 65 percent means that only 35 percent of the potential ability to lift or move is being converted to heat by friction in the pump.

HOW THE PUMP PERFORMS WORK

The hydraulic pump must be driven by a power source **(Figure 2-22)**. The crudest of these would be a hand-operated lever or crank. More common today is either an electric motor or an internal combustion engine. The motor or engine rotates the input shaft of the pump either directly or through a disengagement device such as a clutch. As the input shaft of the pump rotates, the gears, pistons, or vanes move, and as they move hydraulic fluid is moved from the inlet to the outlet. **Chapter 4** will offer detailed information about how each of the many types of pumps operate and the unique characteristics of each type.

TEMPERATURE CHANGES IN THE PUMP

As was previously mentioned, the movement of the fluid through the system generates heat. The pump is attempting to move the hydraulic fluid against a restriction, this generates pressure. As the pressure on a

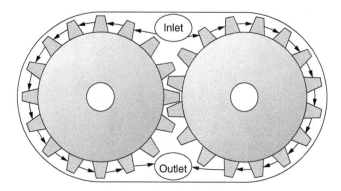

Figure 2-23 The teeth of the gear pump will deploy all between the circumference of the gear and the walls of the pump housing. The gaps between the teeth carry the volume of oil on every rotation much like the paddle wheels of a 19th-century riverboat move through the water to propel the boat forward. The close tolerance of the gears with each other and with the housing can generate a lot of friction. If the pump is getting an adequate supply of flu, this friction is not caused by metal to metal contact but rather by metal to oil and oil to metal contact.

mass, whether it is a gas, a liquid, or a solid increases, it forces the molecules that make up the substance closer together. The closer the molecules are pressed together, the more energy they give up in the form of heat. Friction of the internal moving parts of the pump, friction from the movement of the fluid through the pump, and the pressurizing of the fluid all contribute to increasing the temperature of the hydraulic fluid passing through the pump **(Figure 2-23)**.

LEAKAGES IN THE PUMP

Two categories exist regarding places for the fluid to escape. Fluid escapes externally in the form of leaks. It can also escape internally. Internal escape of the fluid is called bypassing. Bypassing occurs when wear or defect causes fluid being pushed through the system finds it is easier to escape around the internal clearances of the pump gears, pistons, or vanes than it is to move against the operating components and other restrictions of the rest of the system **(Figure 2-24)**.

Bypassing has two basic causes. The first is wear. With excessive wear, tolerances inside the pump increase to the point where bypassing can easily occur. The second cause is instances when restrictions in the system reduce the flow rate to a level far below the volumetric capacity of the pump. When this occurs, the pump attempts to force large quantities of fluid through a low-flow capacity system. The result is a dramatic pressure increase. Eventually it is easier for the fluid to internally bypass the pump than to force its way through the system. This condition leads to

Figure 2-24 High operational hours and damage can cause the pump gears to wear to the point where it is easier for the gears of the pump to move through the oil than it is for the gears to move the oil. This condition decreases the efficiency of the pump by limiting its flow capacity. Eventually the maximum pressure the pump can work against begins to decrease.

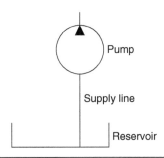

Figure 2-25 Hydraulic oil is supplied to the pump from the reservoir. The line or hose that connects the reservoir to the pump is known as the supply line. This line must be free of kinks and obstructions in order to provide a free-flowing supply of hydraulic oil to the pump and system. A weak or slow operating system has often been diagnosed to a simple kink in the supply hose.

extreme heat from the friction of the fluid attempting to flow. Although systems are designed with **pressure relief valves** to prevent or reduce the likelihood of this situation, when multiple defects occur in a system, catastrophic failures can result, such as defective pumps or sudden and severe leaks **(Figure 2-25)**. It should be noted at this point that the most common cause of multiple failures in a system is contamination.

FLOW IMPEDANCE IN THE PUMP

Two primary causes of reduced flow from the pump are related to the pump itself. One is wear leading to internal bypassing as previously discussed.

Figure 2-26 More complex pumps can have more complex problems. In this variable displacement piston pump, a damaged or binding pivot point, as highlighted by the lower circle, can affect the output of the pump, as well as a defective connection between the control lever and the variable angle swash plate.

The other problem applies to more complex pumps that have internal valves designed to regulate flow or limit pressure or control pressure **(Figure 2-26)**. If this valving is defective, this can also reduce flow.

Keep in mind when troubleshooting a system that the most common causes of low flow in a system are not related to the pump. Restrictions in hoses, lines, valves, and other components are far more likely to cause flow-related and pressure-related issues than in the pump.

FLUID VELOCITY CHANGE IN THE PUMP

The very purpose of the pump is to change the fluid from its relatively static state in the reservoir to the dynamic state of moving through the system to do work. The velocity of the fluid moving through the pump, and therefore through the system, is determined by the rotational speed of the pump. In almost all cases, the speed of the motor or engine driving the pump will determine the speed of the pump and therefore the velocity of the fluid moving through the system **(Figure 2-27)**.

FLUID DAMAGE IN THE PUMP

The pump contributes to damaging the fluid in two significant ways. The first is the already much discussed enemy of heat. There is a greater opportunity for heat to be generated in the pump than there is in

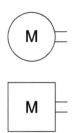

Figure 2-27 These two symbols represent two common power sources for the hydraulic system pump. The upper symbol is an "M" in a circle. This represents an electric motor being used to power a pump. The "M" within a square represents an internal combustion engine, either gas or diesel, or a turbine engine being used to power the pump.

any other component of the system. There are metal components turning and rotating at high speeds. There are high pressures on the outbound side of the pump. Both of these can generate heat.

The second way in which the fluid can be damaged is by metal contamination.

FLUID CONTAMINATION IN THE PUMP

The primary source of contamination from the pump is metal shavings or particles. These result from both normal and abnormal wear of the components. The particles resulting from normal wear are nearly microscopic, and although they are destined to pass through the system once, they will be trapped by the filter as the hydraulic oil returns to the reservoir **(Figure 2-28)**.

Figure 2-28 Contamination can lead to severe damage in all components of a fluid power system. Since the pump receives the hydraulic oil directly from the reservoir, it may be the most vulnerable to this damage. This shows the result of hard particle damage to the pump gears. The wear on the gear limits the flow and pressure capability of the pump. (Note: this pump has been cut open to reveal the damaged gears in place for instructional purposes.)

As the pump ages, or as the pump becomes the victim of contaminated or heat damaged hydraulic fluid, larger pieces of metal can be shed from the pieces inside the pump. This shedding can come from the abrasive action of contamination in the hydraulic oil or from metal to metal contact and battering that result from poor lubrication or from mechanical defects in the pump. Although these contaminates will theoretically pass through the system only once and be trapped by the filter, that single pass can do extreme damage to the components of the system **(see RS 1-2, p. 40)**.

PRESSURE RELIEF VALVE

Purpose of the Pressure Relief Valve

As previously discussed, pressure is the characteristic of fluid in a hydraulic system that provides the "strength" to make a system lift, push, or move. Pressure—excessive pressure—is also the most prolific generator of heat, and heat along with contamination is one of the two most destructive forces at play on a hydraulic machine. Nearly all systems will have a pressure relief valve to limit pressure demanded of the pump. This pressure relief valve is generally set well above the operating pressure of the system. If the system is asked to lift, push, or move too much mass, the increased pressure caused by that request will cause the pressure relief valve to open, which allows fluid to return from the tank. This valve will limit the pressure to a safe level for the pump and other components **(Figure 2-29)**.

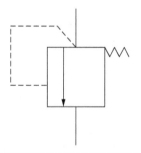

Figure 2-29 The pressure relief valve symbol features a square with an arrow positioned to the side instead of directly in the line of the main fluid flow. This represents the fact that this is a normally closed valve. The zigzag line indicates that the valve is held closed by a spring. The dashed line represents a "pilot" line. When the pressure in the pilot line reaches the valve opening pressure, the valve will open allowing fluid to flow through the main path of the valve. Devices like this are used to ensure the pressure in a system does not rise above safe operating levels.

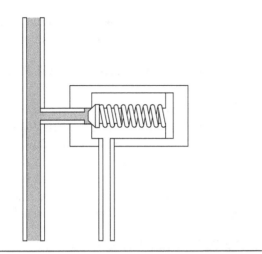

Figure 2-30 This is a representation of a simple pressure relief valve. The hydraulic oil travels parallel to the valve under almost all operating conditions. If the oil in the main path of flow exceeds the valve's calibrating pressure, the valve will be pushed back against the tension of the spring. Fluid can then flow, relieving excess pressure back to the reservoir.

Although the pressure relief valve will be explained more in subsequent chapters, the valve mainly consists of a spool valve and a spring. One end of the spool valve serves as a piston. When pressure on the piston overcomes the spring pressure, the spool valve moves, opening a passage to a return or drain line back to the reservoir.

FLUID FLOW IN THE PRESSURE RELIEF VALVE

The pressure relief valve is typically teed into the main output line from the pump (**Figure 2-30**). This means it is located parallel to the main output line. As fluid flows from the pump to the working components of the system, it also flows toward the closed port of the pressure relief valve. When the fluid reaches the closed port, it stops. As long as the pressure in the working part of the system remains below the opening pressure of the pressure relief valve, the valve will remain closed and all flow will be directed toward the working components of the system. When the pressure as applied to the area of the pressure relief valve overcomes the spring tension, the valve will open and the excess pressure will be relieved back to the reservoir.

HOW THE PRESSURE RELIEF VALVE PERFORMS WORK

No actual work is normally done by the pressure relief valve. Under most operating conditions the valve sits dormant. Fluid passes by having no affect on

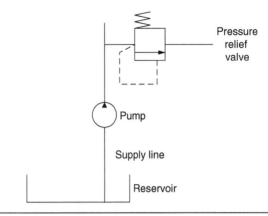

Figure 2-31 In most fluid power systems, the pressure relief valve is located very close to the pump in the outbound line. In fact, because it is the primary device that prevents any part of the system, including the pump, from being damaged by overpressurization, there usually is no more than a short length of hose separating the valve from the pump.

the operation of the valve or on the system (**Figure 2-31**). It is only when the pressure exceeds the relief value that the valve opens returning fluid to the reservoir.

TEMPERATURE CHANGES IN THE PRESSURE RELIEF VALVE

As previously stated, heat is the result of friction. The areas of the system where work is being done are the areas of the system where the greatest amount of heat is generated. Since under most operating conditions the pressure relief valve is passive, the only heat generated is the relatively small amount of heat generated by the fluid passing across the subcomponents of the valve.

LEAKAGES IN THE PRESSURE RELIEF VALVE

The simplicity of the pressure relief valve makes the possibility of external leakage relatively unlikely. Just remember that connection points for hoses and lines are always possible locations for leakage.

Internal leakage however is another story. Time and heat can contribute to metal fatigue on the spring that holds the valve closed. Although the valve itself contributes very little to the total amount of heat in the fluid and system, ambient conditions, reservoir sizing, fluid level, and system cooling efficiency can contribute significantly to the heat to which all of the components are subjected. As the spring weakens, it will eventually allow the pressure and force in the main part of the system to force the relief valve open and allow fluid to escape from the main part of the system back to the reservoir (**Figure 2-32**). This limits the maximum operating pressure of the system and

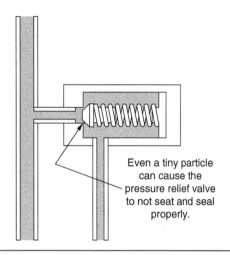

Even a tiny particle
can cause the
pressure relief valve
to not seat and seal
properly.

Figure 2-32 Contamination in the system, especially hard particle contamination, can get stuck in open valves preventing the valve from closing properly. This can cause internal leaks in the system that can result in significant deterioration in the performance of the system.

therefore limits the maximum lifting or pushing capacity of the system.

The seal between the relief valve itself and the housing of the valve assembly is usually metal to metal. Because this valve in most system designs rarely cycles, that is to say rarely opens and closes, the wear factor on the valve is minimal. Nonetheless, all metal components are subject to wear. Wear can cause an imperfect seal, allowing an internal system leak back to the reservoir.

Perhaps the highest risk for an internal leak past the pressure relief valve comes from internal contamination. In contaminated or poorly maintained systems, contamination such as dirt, metal fragments, and other particles can and will settle in areas where the flow of the fluid creates an eddy. Although the systems and system components are designed to minimize these eddies, they still exist. When the pressure relief valve performs its normal function and opens, there is an opportunity for contamination to move into the valve and prevent the valve from closing when the pressure in the system allows the relief valve to close. This would cause an internal leak that could limit the lifting or pushing power of the system.

FLOW IMPEDANCE IN THE PRESSURE RELIEF VALVE

There is little in the pressure relief valve to impede flow. In most pressure relief valves, the flow of the fluid is parallel to functional components of the valve. Therefore, if the valve has been selected correctly by the designer of the system then there should be virtually

no impedance to the flow of the fluid through the pressure relief valve when the valve is not relieving.

FLUID VELOCITY CHANGES IN THE PRESSURE RELIEF VALVE

Changes in velocity always occur with a change in diameter or a change in conduit material. These changes cause a change in the resistance to flow; this causes both a pressure change and a change in fluid velocity. Assuming the pressure relief valve has been properly chosen by the system designer, there should be no change in velocity as the fluid flows through the valve.

FLUID DAMAGE IN THE PRESSURE RELIEF VALVE

There is very little that the pressure relief valve can do to damage the fluid. Unless a defect exists in the valve, there is no source of heat to cause damage to the hydraulic fluid.

FLUID CONTAMINATION IN THE PRESSURE RELIEF VALVE

Since the pressure relief valve is in the sealed side of the system, no contamination can enter the system. Damage to the valve can result from contamination from other sources. This contamination could then cause damage to the pressure relief valve. The damage to the pressure relief valve could then further contaminate the system. This contamination would be the secondary, not primary, cause of damage (**see RS 1-3, p. 40**).

DIRECTIONAL CONTROL VALVE

Purpose of the Directional Control Valve

A wide range of control valves can be installed in a system. For our tour through the system, we will discuss one of many possible **directional control valves (Figure 2-33)**. The directional control valve is fundamental to the operation of most hydraulic systems. The purpose of the valve is to route the fluid in the proper direction through the actuator to lift or push. There are typically, though not always, three positions for the valve. The center position is usually the idle position. In this position, the valve either blocks the flow of fluid or routes the flow back to the reservoir, bypassing the actuator. Each of the two non-middle positions routes the fluid through the actuator in opposing directions. In one direction the fluid would be routed through the actuator to lift a load or push a load forward. Move the valve to the other non-middle position

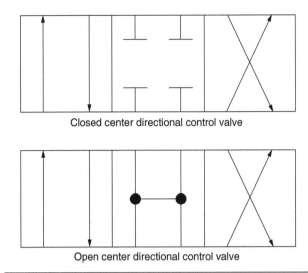

Closed center directional control valve

Open center directional control valve

Figure 2-33 The valve with which most system operators will have the most direct contact is the directional control valve. The top symbol represents a closed center directional control valve. When these valves are in the neutral or center position, fluid cannot flow in any direction through the valve. The lower valve is an open center valve. When these directional control valves are in the center position, the two sides of the system—the pump and actuator sides—are separated. However, fluid can still flow freely through the valve on the pump side and fluid can flow freely on the actuator side, hydraulic oil just cannot cross from the pump side to the actuator side.

and the fluid will flow the opposite way through the system. The load that was lifted is now lowered, or the load that was pushed forward will now be pushed in reverse.

FLUID FLOW IN THE DIRECTIONAL CONTROL VALVE

In one common form of the directional control valve, there is one inlet port and one outlet port on each side of the valve. Flow from the pump enters the inlet port. If the valve is in the center, or neutral, position, the oil is routed directly back to the reservoir and the pressure between the pump and the directional control valve is minimal. Valves that feature this free return of the fluid to the reservoir in the neutral position are called *open center valves*.

The second common type of valve also has one inlet and one outlet port on each side of the valve, but when the valve is in the center, or neutral, position the fluid is *deadheaded* or blocked from flowing back to the reservoir or to anywhere else. In these systems, the pressure between the pump and the directional control valve rises to the set pressure for the pressure relief valve, or a secondary pressure relief valve set a little

above maximum operating pressure for the system when the system is operating.

HOW THE DIRECTIONAL CONTROL VALVE PERFORMS WORK

For the purpose of this walkthrough of a basic hydraulic system, the inbound side of the directional control valve has an inlet port and an outlet port. The outbound side of the valve also has two ports. When the spool in the valve is shifted in one direction (for example, to the left) the fluid flows out one port ("Port A," for instance) and returns from the controlled device through "Port B." If in this left position Port A is connected to the extend side of a cylinder, fluid will pass through Port A to the extend side of the cylinder and the piston of the cylinder will begin to move to extend the cylinder. Assuming the cylinder as a double acting cylinder, the fluid in the retract side of the cylinder must move out of the cylinder and back to the reservoir in order for the piston to extend. This fluid will pass out of the retract side of the cylinder and through Port B, then though the outlet port on the inbound side of the valve and back to the reservoir.

When the directional control valve is moved to the right side, the flow through the valve from the pump shifts to Port B and therefore to the retract side of the cylinder **(Figure 2-34)**. The fluid in the extend side of the cylinder must then return back to the reservoir though Port A, which is now connected through the valve to the port connected to the return line to the reservoir. The piston in the cylinder therefore retracts. The fluid will pass out the extend side of the cylinder, through Port A, and then through the outlet port on the inbound side of the valve and back to the reservoir **(Figure 2-35)**.

The end result is that when the valve is in the center position the cylinder is not extending or retracting. Move the valve to the left and the piston of the cylinder will extend. Move the valve to the right side and the piston will retract.

This same type of valve could be used to stop and start a hydraulic motor. Depending on whether the valve was in the left or the right position, the motor would turn either direction.

TEMPERATURE CHANGES IN THE DIRECTIONAL CONTROL VALVE

Because no real work is taking place in the directional control valve there is little opportunity for heat to be generated. The only heat is that resulting from the friction of the hydraulic oil passing through some of the

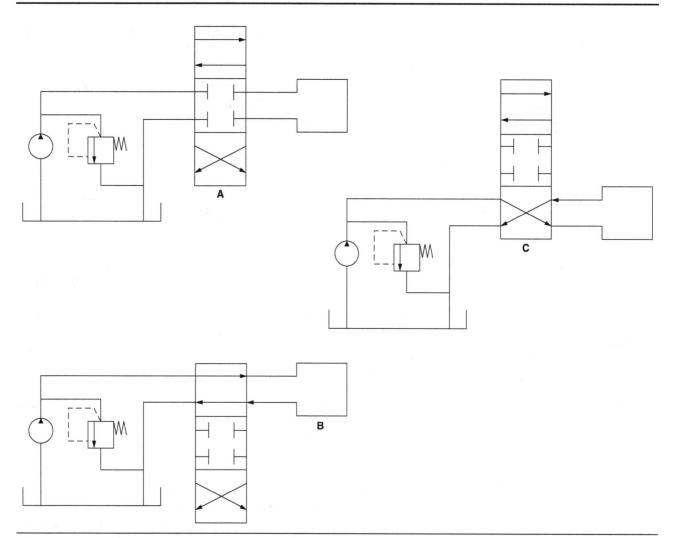

Figure 2-34 Drawing "A" shows a closed center directional control valve in the center position. The fluid from the pump can only flow up to the directional control valve and then it is stopped. The closed center of the valve amounts to a complete blockage to flow. To prevent damage to the pump or other parts of the system, the pressure relief valve must open. In the "B" drawing, the valve has been moved to the straight through flow across the valve. The pressure relief valve can close and fluid can flow in a forward direction to or through the actuator. In drawing "C," the valve has been moved to the cross flow position. The fluid will now flow to or through the actuator in a reverse direction.

small openings in the valve. In a properly designed system, where the valve has been chosen with a properly specified flow rate even the heat from this friction will be negligible. Also there is almost no opportunity for heat in the oil to be dissipated. The extremely low surface-area-to-flow-rate ratio prevents the heat from escaping the hydraulic fluid **(Figure 2-36)**.

LEAKAGES IN THE DIRECTIONAL CONTROL VALVE

Barring catastrophic failures, the only place external leaks are likely to occur is at the fittings connecting the hydraulic hoses and lines to the directional control valve. It is far more likely for a directional control valve to develop an internal leak. Wear and hard particle

contamination can cause wear between the valve spool and the valve body. This can lead flow through the inlet port to leak directly to the outlet port **(Figure 2-37)**.

FLOW IMPEDANCE IN THE DIRECTIONAL CONTROL VALVE

Other than contamination, the only thing that impedes flow related to the directional control valve is improper sizing of the valve. That said, it should be noted that there is an impeding of flow inherent to almost every component in a hydraulic system. What the system designer or technician needs to remember is to size the directional control valve to match or preferably exceed the maximum flow rate of the operating system.

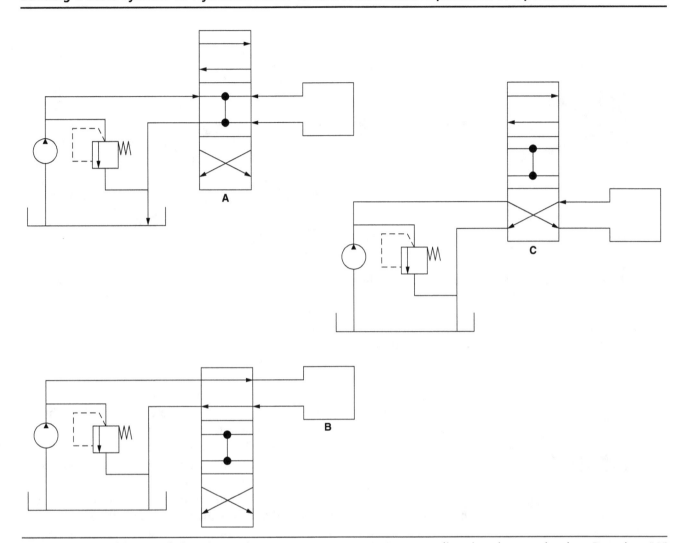

Figure 2-35 In this set of drawings, the system uses an open center directional control valve. Drawing "A" shows the valve in the center position. On the pump side of the system, fluid is flowing freely from the pump through the directional control valve and back to the reservoir. The pressure in the pump side of the system is either minimal or zero. On the actuator side of the system, fluid can move in a free manner back and forth to or through the actuator. The "B" drawing and the "C" drawing show that the behavior of the fluid in the other two positions is no different whether the valve is open center or closed center. (Note that many types of directional control valves will be discussed thoroughly in subsequent chapters.)

FLUID VELOCITY CHANGES IN THE DIRECTIONAL CONTROL VALVE

There is nothing in the directional control valve that inherently has a large affect on the velocity of the fluid. Just like its affect on flow rate, if the valve has been sized properly the flow rate of the system will not be affected and therefore the velocity of the fluid will not be affected.

FLUID DAMAGE IN THE DIRECTIONAL CONTROL VALVE

With little in the directional control valve to add contamination or heat, there is nothing to damage the fluid. The valve itself however can be significantly

damaged by contamination of the fluid from other sources (see RS 1-4, p. 41).

CYLINDER

Many types of actuators exist, given the great variety of hydraulic systems worldwide. However, they all boil down to some type of cylinder or some type of motor. The simplest is a cylinder (**Figure 2-38**). There are many types of cylinders that vary in complexity. For the discussion in this chapter, one of the simpler designs will be used, but not the simplest.

The actuator chosen is a double acting cylinder. This means that hydraulic fluid under pressure caused by a flow from the pump is used to both extend and

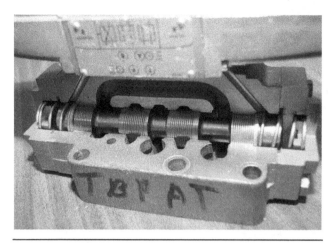

Figure 2-36 This is a cutaway of a directional control valve. The spool that runs through the center of the valve is moved back and forth to direct the flow of hydraulic oil through the system. The closely machined spool seals against very high pressures and yet must slide freely into its various positions.

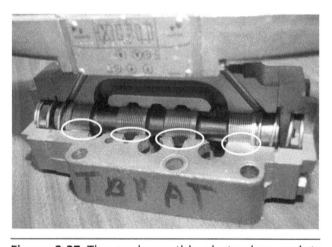

Figure 2-37 The ovals on this photo show points where the valve must seal against pressures as high as several thousand pounds per square inch. These high pressures are sealed by closely toleranced machined surfaces between the spool and the valve body. If the spool valves become damaged or even slighted abraded, they can fail to seal properly, causing erratic operation or no operation of the fluid power system.

retract the cylinder. Other types and designs will be discussed in later chapters.

Purpose of the Cylinder

If the pump is the heart of a hydraulic system, then the actuators are the muscles of the arms, legs, fingers, and toes. In determining how a system operates, which is essential in troubleshooting a system, it is important to remember that a hydraulic cylinder cannot pull, it can only push. A **hydraulic cylinder** consists of a

Figure 2-38 The cylinder is probably the most common fluid power actuator. It is a machine designed to emulate and amplify the action of muscles in the human body. In much the same way muscles work in pairs to most arm, leg, and other joints, the cylinder moves the mechanical appendages of machinery.

cylindrical tube, a piston, and a rod (or rods) attached to the pistons. In the example used here, the cylinder has a single piston designed to seal fluid on either side of the piston. There is a rod connected to only one side of the piston. The directional control valve controls on which side of the piston the hydraulic fluid flows. When the fluid flows on the rod side of the piston, the piston (and therefore the rod) retracts. When it flows to the other non-rod side of the piston, also known as the blind side, the rod is extended. Anything attached to the rod or in front of the rod is then pushed.

FLUID FLOW IN THE CYLINDER

When the directional control valve is moved into the position that permits fluid to flow into the extend side of the cylinder. In a single rod cylinder, such as the example cylinder, this is the blind side or non-rod side of the piston. As the non-compressible fluid flows, it forces the piston to move. The piston and the force of the fluid acting on the piston combine with the load on the system to increase system pressure between the pump and the piston of the cylinder. Increasing the load on the rod increases the pressure. In order for the force applied by the pressure on the piston to actually move that piston, several resistances must be overcome. There is the resistance of the piston and rod in the cylinder; it must also overcome the force of gravity and the inertia of the object it is attempting to move. Keep in mind that part of the load is the force required to push the fluid from the rod side of the piston back to the reservoir **(Figure 2-39)**. When the directional control valve is moved to the opposite position, the

Figure 2-39 In this photo, two hoses are used to deliver fluid from the directional control valve. In this case, when extending the cylinder the fluid flows to the cylinder through "A" and returns from the cylinder through "B." When the cylinder is to be retracted, the direction of the flow is reversed. The oval on the photo highlights a place where a hose has been rubbing against the articulating structure of the machine. Care must be taken to ensure that the hoses are positioned to prevent such damage. This should have been prevented by repositioning the hose during routine maintenance long ago.

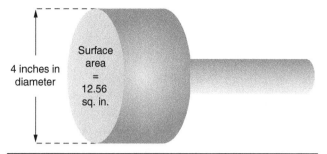

Figure 2-40 A typical fluid power piston will have a rod attached to one end. The opposite side will not have a rod; this is called the "blind" side of the piston. The surface area of the blind side of the piston is calculated by multiplying the radius of the piston—in this case, 2 inches—by itself, or squaring the radius. The result is then multiplied by pi, or 3.14, to calculate the surface area. $2 \times 2 \times 3.14$ equals 12.56. This is the surface area of the blind side of the piston.

fluid flows to the rod side of the piston and the piston is retracted. The same issue of gravity, friction, and inertia must also be overcome to retract the piston.

HOW THE CYLINDER PERFORMS WORK

As the fluid pressure builds on the blind, or extend, side of the piston, it begins to exert a force on that piston. If the piston is 4 inches in diameter, it has a surface area of 12.56 inches. To calculate the area of a circular piston, multiply the radius of the face of the piston times itself, or in other words, square the radius. Remember that the radius is half the diameter. This is 2 times 2 in the example, which is 4. Multiply the result of squaring the radius by π (pi). Pi is a mathematical constant of approximately 3.14 that is defined as the ratio of a circle's circumference to its diameter. Multiplying 3.14 times 4 yields a result of 12.56 square inches (Figure 2-40). When 1 pound per square inch (psi) is applied evenly across the surface of the piston, the force will be 12.56 pounds. Note that Pascal's Law says the pressure will always be applied evenly across the face of the piston. When the pressure rises to 10 psi, the force applied to the piston, and therefore to the load, will be 125.6 pounds. As the

pressure continues to increase, the force also increases proportionally.

When the force applied by the fluid is great enough to overcome the resistance of the load to movement, the load will begin to move. As the piston begins to move, the fluid located on the retract side of the piston must flow through the return line back to the reservoir.

When it is time to retract the rod, something rather unexpected happens. Since the rod is attached to this side of the piston, the surface area of the piston occupied by the rod cannot exert a force. Using the same 4-inch piston, its 12.56-square-inch surface area is reduced by the area of the rod. If the cylinder has a 2-inch rod, the surface area occupied by the rod is 1 times 1 times 3.14. This is 3.14 square inches. By subtracting this result from the area of the piston, we get an effective surface area of only 12.56 minus 3.14 square inches, or 9.42 square inches (Figure 2-41). As a result, when 100 psi is applied to the retract side of the piston, the force applied to the piston is only 942 pounds. Assuming the force required for moving the piston, rod, and load is the same in both directions, and often it is not, then higher pressures are required to retract than are required to extend.

TEMPERATURE CHANGES IN THE CYLINDER

The piston, rod, cylinder, and load are the main resistance to the flow of the fluid. It is their resistance to movement that causes the hydraulic fluid in the system to pressurize. Along with friction, pressure is the primary cause of heat in the system. As the pressure increases against the resistance to movement, the pressure in the working side of the system

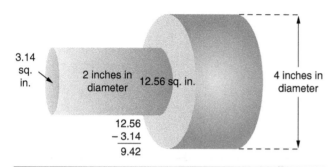

3.14 sq. in.

2 inches in diameter   12.56 sq. in.

4 inches in diameter

12.56
− 3.14
9.42

Figure 2-41 The effective surface area of the rod side of the piston is the surface area of the piston less the surface area of the rod. The rod is 2 inches in diameter; this means it has a radius of 1 inch. 1 times 1 is 1. Pi times 1 is 3.14. Therefore, the area of the surface of the piston displaced by the rod is 3.14 square inches. Thus, the effective area of the rod side of the piston is 12.76 square inches less 3.14 square inches, or 9.42 square inches.

increases. Thus, as the pressure increases, so does the temperature.

It should be noted that when the piston seals begin to leak and allow hydraulic fluid to pass between the extend side and retract side of the piston (or vice versa) the friction of the bypassing fluid also generates heat.

LEAKAGES IN THE CYLINDER

The cylinder used in this example system has two primary places for leaks. One is an external leak where the rod enters the cylinder. This is easy to diagnose since oil will be evident around the shaft where the rod enters the cylinder **(Figure 2-42)**.

Two is less obvious; it is the internal leakage of the hydraulic fluid bypassing the piston seals. This results in excessively high-temperature return fluid. The high-temperature return fluid is often accompanied by chatter or vibration as the fluid bypasses the seals. One of the best ways to diagnose this condition is with the use of an infrared sensing thermometer.

FLOW IMPEDANCE IN THE CYLINDER

Since the cylinder is the component of the hydraulic system actually doing the work, it is only reasonable to assume that it will cause a resistance to flow. Beyond this normal resistance there can be additional resistance due to damaged piston seals, due to scoring or damage in the cylinder walls. A bent rod can also provide additional resistance to flow.

FLUID VELOCITY CHANGES IN THE CYLINDER

The actuator, in this case the cylinder, is the place that has the greatest resistance to flow. If the effect of

Figure 2-42 Hydraulic oil can escape from a cylinder in two chief places. Actually, the first is not the cylinder itself but rather the process in carrying the fluid to and from the cylinder. This is highlighted by the upper oval in the photo. There is an obvious fluid leak in one of the hoses. The other likely place is at the seal where the rod enters the cylinder. In this example the lower oval shows the location. There is obvious discoloration around the body of the cylinder. This type of discoloration is only superficial staining of the paint on the cylinder.

the fluid acting on the surface of the piston is unable to develop enough force, the piston cannot move. The fluid being pushed by the pump must flow somewhere. Initially the flow will compress the fluid. Since the compressibility of the fluid is only one half of one percent of its volume per thousand pounds per square inch (0.5 percent per 1,000 psi) the fluid will compress rapidly to its maximum compression and the pressure will rise rapidly. If the load is so great on the rod that it refuses to move, the fluid must find a way to escape. This is where the pressure relief valve begins to act. The building pressure creates a force against the piston of the pressure relief valve, which will quickly overcome the spring tension acting on the other side of the piston and the valve will open. If the valve does not open due to contamination, damage, or defect, the fluid will still have to flow. At this point in the example simple circuit, the fluid will either force its way out of the weakest line, junction, or fitting or bypass the internal components of the pump. The broken line will create a mess; the internal bypass will generate extreme heat and rapidly damage the pump.

If the piston in the cylinder moves, then the pressure between the piston and the pump will rise only to the point where enough force is created to move the load. As the difficulty in moving the load fluctuates, the

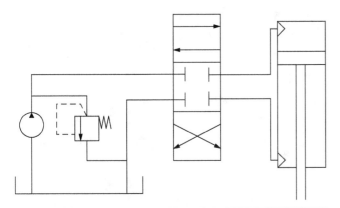

Figure 2-43 When there is no load on the cylinder, there is very little resistance to the movement of the piston and rod. If the directional control valve were moved to either the straight flow or cross flow position, the pressure in the system would be minimal—only the very light pressure required to move the mass of the piston and rod.

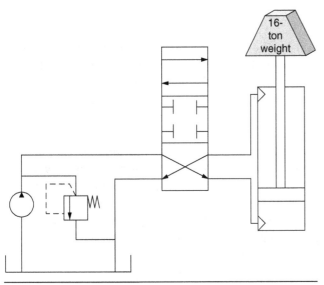

Figure 2-44 Add a large mass or obstruction to the movement of the rod and the pressure in the system will increase dramatically. As the resistance to movement increases (as with the 16-ton weight in the drawing), the pressure in the system will build to the point where the weight begins to move. If the weight is too great, then the pressure relief valve should open to protect the pump and other components. If the pressure relief valve fails to open, major damage can occur.

force required to move the load will also fluctuate, as will the pressure. When the piston in the cylinder reaches the point of full extension, the piston will stop moving. Pressure will rise **(Figure 2-43)**. The pressure relief valve will have to open or the pump will begin to bypass internally, meaning heat will be generated and damage will occur to the system **(Figure 2-44)**.

It should be noted that this example is a very simple system and that in most systems redundant safeguards exist to protect the pump. In addition to the pressure relief valve described earlier, there may be a second pressure relief valve set to a slightly lower pressure which relieves pressure when the piston reaches the end of its travel. This second valve may be a separate component or may be located inside the pump. It does not matter if it is upstream or downstream of the system pressure relief valve since the pressure will be the same at all points between the pump and the piston.

FLUID DAMAGES IN THE CYLINDER

Since the actuator—in this case, the cylinder—is the component doing the actual work in the system, it is the place where heat is most likely to be generated. Actually, the heat is generated from the pump to the piston by the pressure required to move the piston against the load. Heat is one of the greatest causes of damage and deterioration to the fluid.

It should be noted that as the size of the surface area of the piston increases, the force generated by a lower pressure will also increase. A large piston will move the same load at a lower pressure than a small piston. The trade-off is that a greater volume of fluid is required to move a larger piston. If the pump volume and volume carrying capacity of the lines and hoses are unchanged, the larger piston will tend to move slower. The lower operating pressure does mean that less heat will be generated and there will be a lower tendency to damage the hydraulic fluid.

FLUID CONTAMINATION IN THE CYLINDER

Typically, there is little about the cylinder to contaminate the fluid. As the piston and seals deteriorate, they may begin to shed seal material, metal shavings, and rust into the hydraulic fluid. These contaminates should be removed from the fluid by the filter located in the return line of the reservoir **(Figure 2-45)**. Excess contaminates may damage the filter and allow these particles to spread throughout the system **(see RS 1-5, p. 41)**.

HYDRAULIC MOTOR

Purpose of the Hydraulic Motor

Many systems do not utilize cylinders to do work, but rather they use hydraulic motors to move their loads. Just like with cylinders, many types of hydraulic motors exist. This example uses a simple bidirectional motor. This means that when fluid flows in "Port A" and out "Port B," the motor turns one direction and

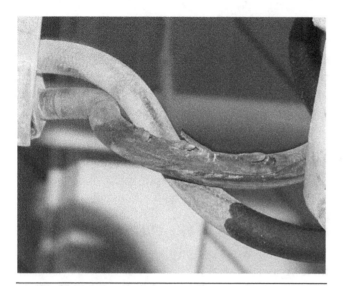

Figure 2-45 Many hoses in a fluid power system are sheathed to protect them from rubbing on the mechanical components of the equipment. Improper maintenance or failure to protect the hoses from being chafed by moving components can damage the hose and possibly cause system contamination and massive leaks.

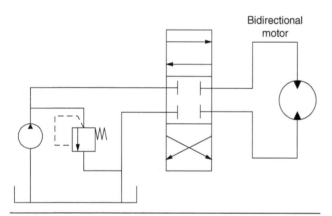

Figure 2-46 In addition to cylinders, another common actuator is the hydraulic motor. Fluid flows through the motor rotating a set of gears or vanes, or perhaps causing a set of pistons to reciprocate. It is no accident that the symbol for a motor is almost identical to the symbol for a pump. Many motors have internal components that look identical to a pump.

when it flows in "Port B" and out "Port A," the motor turns the other direction. Motors are used for devices such as crane turntables and winches. Many types of motors exist, just as many types of pumps do **(Figure 2-46)**.

FLUID FLOW IN THE HYDRAULIC MOTOR

Hydraulic fluid is forced through the hydraulic motor as a flow created by the pump **(Figure 2-47)**. In much the same way rotating or reciprocating gears or pistons in the pump force the fluid to flow through the

Figure 2-47 This hydraulic motor is a simple gear motor. Notice that the return line slides onto a simple tube fitting designed to accept a rubber hose and a clamp. This indicates it is a unidirectional motor. If it were bidirectional, then both hoses would have to be attached to the motor with high-pressure fittings.

system, the system's flowing fluid causes the gears or pistons in the motor to rotate or reciprocate. If there is little or no load on the motor, the force needed to make the motor rotate is minimal. As the load increases, the force required to make the motor rotate also increases. The increasing demand for force causes the pressure to rise. As with the cylinder, the amount of pressure available to rotate the pump will determine how much the pump will lift. Just like with the cylinder, the physical dimensions of the motor will determine the maximum speed, load ability per psi, and overall performance. That discussion is reserved for a subsequent chapter.

HOW THE HYDRAULIC MOTOR PERFORMS WORK

A set of gears or pistons is subjected to the flow created by the pump **(Figure 2-48)**. Just like in a cylinder, if the load on the motor is minimal, then the pressure in the system will be minimal. As the load increases, the resistance to rotation of the motor also increases. The resistance to flow causes the pressure from the pump to increase. As the load is increased, at some point the load will become greater than the system's design parameters. At that point, if the system is designed correctly, the pressure relief valve will open and the motor will stop lifting or pushing the load. If the motor stalls before the pressure relief valve opens, the flow from the pump will still want to continue. Either the motor or the pump must bypass or leak internally.

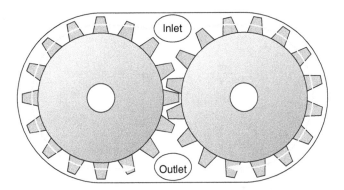

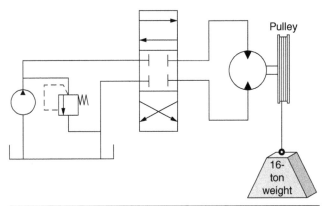

Figure 2-48 In a simple had rolling motor, the oil will enter from one passageway that serves as an outlet. It is drawn around the outer circumference of the gears between the gear teeth and the pump housing. This action, like with the pump, moves the teeth of the gears in much the same way as the paddles on a medieval water wheel is rotated by the water passing over or under the wheel. One of the gears will be attached to the output shaft of the motor.

TEMPERATURE CHANGES IN THE HYDRAULIC MOTOR

When pressure increases, the temperature rises. Also, should the gears or pistons wear enough to allow fluid to bypass the gears or pistons, the bypassing fluid will generate friction as it seeps past the worn or defective components. The heat will cause friction that in turn will generate heat. In addition to the friction caused by the fluid flow, there is also some friction and therefore heat created by the moving parts of the motor.

LEAKAGES IN THE HYDRAULIC MOTOR

Just as in the pump and many other components, the fluid can escape externally, in which case a leak occurs and puddles of hydraulic fluid can be found on the ground. There can also be an internal leak. When the pressure attempting to rotate the motor becomes too high, hydraulic oil can be forced past the gears and pistons forcing the fluid to return to the reservoir. As the components inside the hydraulic motor deteriorate due to age and wear, the pressure at which this bypass can occur will drop, eventually rendering the system incapable of lifting its maximum load.

FLOW IMPEDANCE IN THE HYDRAULIC MOTOR

The flow is impeded by the attempt to lift or move a load with the motor. Since the physical characteristics of the example motor are fixed, the only variable that can change to affect the load is pressure. The greater the load on the motor, the greater is the resistance to flow; the greater the resistance to flow, the greater the pressure in the system. Due to the resistance to flow created by the load on the motor,

Figure 2-49 When there is a light load on the motor, only a small pressure is required in the system to make the motor move. As the load increases the pressure required to make the motor move will also increase. Many motors are attached to pulleys or gears. The larger that pulley or gear, the less effort it will take to lift or move a given load. Therefore, the larger the gear or pulley, the lower the operating pressure for a given load.

pressure rises **(Figure 2-49)**. When the load becomes too great, the pressure relief valve or another pressure regulating valve will open and divert flow back to the reservoir to limit the pressure.

A damaged motor can be more difficult to turn than a motor that is not damaged. Anything that impedes the rotation of the motor, including damage to the motor itself, will impede flow, which in turn will tend to increase operating pressures and therefore operating temperatures.

FLUID VELOCITY CHANGES IN THE HYDRAULIC MOTOR

As with the cylinder-based system, in a hydraulic motor–based system, the motor is the point with the greatest resistance to flow in the system. If the effect of the fluid acting on the gears and pistons of the motor is unable to develop enough force, then the motor cannot rotate. Like with the cylinder, the fluid being pushed by the pump must flow somewhere. Again in the motor, just as with the cylinder, the flow will compress the fluid. Since the compressibility of the fluid is only one half of one percent of its volume per thousand pounds per square inch (0.5 percent per 1,000 psi), the fluid will compress rapidly to its maximum compression and pressure will rise rapidly. If the load is so great on the motor that it refuses to rotate, the fluid must find a way to escape. The pressure relief valve must begin to react.

If the load requires a force that is created by a pressure lower than the relief valve pressure, the motor will begin to rotate. Since the pump in the example system supplies a steady flow, the motor must turn at a rate that is proportional to flow. Just as with the cylinder, the flow rate will determine the velocity of the fluid, and the velocity of the fluid will determine the rotational velocity of the motor, which in turn will determine the velocity at which the motor moves the load.

FLUID DAMAGE IN THE HYDRAULIC MOTOR

The motor in this example is our actuator. Just like the previously discussed cylinder, it operates by creating a restriction to flow. The greater the load on the motor, the greater is the pressure in the system. The greater the pressure, the greater is the heat. The amount of heat, and therefore the effects of that heat on the system, can be mitigated by hydraulic fluid coolers, and even more effectively by using a physically larger motor to move a given load.

FLUID CONTAMINATION IN THE HYDRAULIC MOTOR

Nothing in a hydraulic motor can contaminate the fluid until the motor begins to deteriorate. As the gears or pistons and the seals deteriorate, they may begin to shed seal material, metal shavings, and rust into the hydraulic fluid. These contaminates should be removed from the fluid by the filter located in the return line of the reservoir. Excess contaminates may damage the filter and allow these particles to spread throughout the system (see RS 1-6 and RS 1-7, p. 42).

RETURN LINE

The return line completes the hydraulic fluid's journey back to the reservoir. The fluid returning from the actuator passes through the return ports of the directional control valve and into the return line.

Purpose of the Return Line

The purpose of the return line is quite simple: It is the path or conduit through which the fluid passes in returning back to the reservoir (**Figure 2-50**). Perhaps the most critical issue about the return line is that it does not increase restriction or resistance to the flow of the hydraulic fluid. A restriction in the return line, like a restriction at any other point in the system, will cause pressure to rise upstream of the restriction and reduce the flow.

The rise in pressure ahead of the restriction in the return line means that the pressure differential across the actuator, either the cylinder or the motor, is reduced. If there was a restriction that caused pressure between the actuator and the restriction to be 100 psi, then the lifting or moving power of the actuator will be

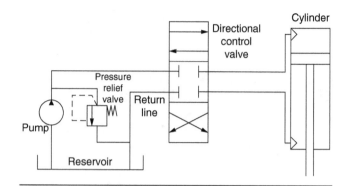

Figure 2-50 The return line brings the hydraulic oil back to the reservoir. Since the pressures are typically low in the return line, the material used in the line can be of lightweight materials.

reduced by the amount of force that would have been created by that 100 psi.

The restriction in the return line will impede flow, particularly in a cylinder actuator system. This can slow the operation of the cylinder every bit as much as the same restriction in an inbound line to the cylinder. Simply, if fluid cannot flow out one side of the cylinder, it will not be able to flow into the other side of the cylinder.

FLUID FLOW IN THE RETURN LINE

Fluid flows into the return line from the directional control valve. The hydraulic fluid pressure is very low in the return line, very little heat is generated, and in many cases the return line begins the cooling process for the fluid.

HOW THE RETURN LINE PERFORMS WORK

The return line is a simple low-pressure passageway from the working part of the system back to the reservoir. The hose or line is usually made of a relatively low-grade material since it does not handle either high pressures or high temperature (**Figure 2-51**). In spite of this, it is usually a bad idea to randomly choose a material for replacement hoses. Although the return line does not require as much strength as the line from the pump to the actuator, it does need to be compatible with the type of hydraulic fluid being used. Incompatible hose material can deteriorate and shed damaging material into the reservoir.

TEMPERATURE CHANGES IN THE RETURN LINE

The pressure is typically low in the return line. As a result, there is an opportunity for temperatures to begin to drop. The cooling capacity of the return line is generally very limited because of the low surface area of the hose.

Figure 2-51 In this system, the highlighted return line is connected to the reservoir with a lightweight and relatively weak compression fitting. Rigid piping like this is typically only used in locations where the piping and reservoir can both be rigidly mounted to a framework or panel.

LEAKAGES IN THE RETURN LINE

Return lines are a common source of leaks. There may be many reasons for this. The relatively low quality of the hose material is certainly one reason, another is the fact that all of the contaminants in the system end up in the return line as they head toward the system filter. Water, oil, and other contaminants can lead to deterioration of the return line hose material.

FLOW IMPEDANCE IN THE RETURN LINE

Flow through the return line is as important as the ability of the fluid to flow through any other line in the system. Deteriorated hose material, kinks, and foreign objects restricting the hose can reduce flow in the system **(Figure 2-52)**. If the fluid cannot return to the reservoir, it will keep fluid from passing out of the cylinder or hydraulic motor. If fluid cannot pass out of the actuator, then it cannot move into the actuator and therefore it cannot function at the proper speed. Severe restrictions in the return line can cause enough pressure to build up to cause it to burst. Both cylinders and motors depend on a pressure differential between the inlet and the outlet to operate. A restriction in the return line holds back this pressure and therefore reduces this critical pressure differential. Flow is reduced and therefore the speed of operation. The pressure differential is reduced and therefore the lifting and pushing ability or strength of the system is reduced.

FLUID VELOCITY CHANGES IN THE RETURN LINE

In most cases, the return line will be one of the largest hoses in the system. In complex systems there

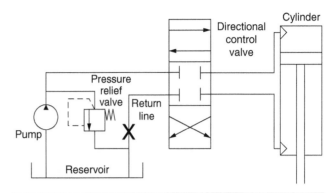

Figure 2-52 A restricted return line (indicated by the "X") will cause the pressure ahead of the restriction to increase. Additionally, this restriction will impede flow throughout the system. The speed at which the system operates will be limited by the amount of fluid that can pass through the restriction. The restriction will determine the flow rate in the system. Also the power in the system, the ability to lift or push, will be limited to the pressure difference between the actuator inlet line and the pressure that is being held back by the restriction. The system will be slow and weakened.

may be many small return lines coming from different areas of the system. Whether there is one large line or several small lines, the total capacity of the return lines must be in excess of the maximum flow rate of the pump. If the capacity of these lines is less than that of the pump, the system will operate with reduced efficiency.

FLUID DAMAGE IN THE RETURN LINE

While traveling through the return line, the hydraulic fluid is under very little stress and no heat is being added. As a result, it is unlikely that the fluid could be damaged while traveling through the return line.

FLUID CONTAMINATION IN THE RETURN LINE

The inner walls of hoses that form the return line are designed to resist deterioration and damage. In the course of repairs and maintenance, if the return line hoses are replaced with improper hoses, these hoses can begin to deteriorate and spread contamination to the filter and eventually to the entire system.

FILTER

Even though the return line completes the journey of the hydraulic fluid through the system, in most systems there is still one last component through which

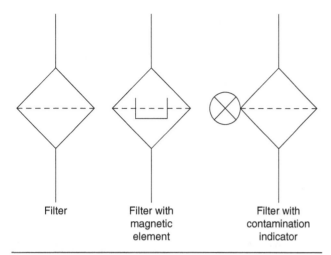

Filter   Filter with magnetic element   Filter with contamination indicator

Figure 2-53 The filter is typically located at the end of the return line. On the left is the symbol for a standard filter. The middle symbol indicates that the filter has a magnetic element to attract and hold metallic particles that have made it through the hydraulic system or have been shed by the system. The symbol on the right features an indicator that informs the operator or technician that the filter is contaminated and needs to be replaced.

the fluid passes as it enters the reservoir. The filter is placed at the end of the return line or lines so that any particulate contamination that may have been picked up by the oil as it passed through the system may be filtered out **(Figure 2-53)**.

Purpose of the Filter

Hydraulic systems have very closely toleranced machining. In many places throughout a typical system, there will be metal-to-metal contact with several hundreds or even thousands of psi of oil pressure being exerted on the point of contact. Very tiny particles, measuring only a few microns across, can wreak havoc on the machining and the sealing ability of these precision parts. The filter is designed to stop these tiny particles from returning to the reservoir. As the system runs, contaminates can eventually restrict the filter. Pressure begins to build behind the filter. If the restriction increases, the pressure can build to the point where the filter can fragment. At that point, there are not just particle contaminants being spread through the system, but also fragments of the filter material itself spreading through the system.

FLUID FLOW IN THE FILTER

Hydraulic fluid enters the filter through an inlet port. From there it passes through the filtration media. Two common filtration materials are cellulose and

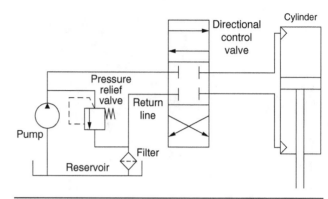

Figure 2-54 The filter is located in the return line just before or just after it enters the reservoir. The filter completes the hydraulic fluid's journey through the system.

fiberglass. When particles being carried by the fluid impact the filter media, they are stopped while the fluid continues to pass by, destined for the reservoir **(Figure 2-54)**.

HOW THE FILTER PERFORMS WORK

There is no work done by the filter other than removing contamination. As mentioned earlier, a restricted filter can impede the operation of both hydraulic cylinders and hydraulic motors.

TEMPERATURE CHANGES IN THE FILTER

Under all operating conditions, when the hydraulic fluid enters the reservoir, the velocity decreases because the internal volume and the cross-section of the filter is larger than the lines bringing the fluid from the **directional control valve** and other sources. This slowing of the fluid gives it the opportunity to cool.

LEAKAGES IN THE FILTER

A filter can have many physical configurations. Some configurations will allow the fluid to escape from the system in the form of an external leak, while others will not **(Figure 2-55)**. There is actually more of a chance of internal leaking than external. Many filters have a bypass design. These filters bypass fluids directly back to the reservoir and thus "bypass" the filtration media. This occurs when the media is so contaminated that it is causing the pressure in the return line to rise, as well as the impeding fluid flow, to the point where the system may operate slowly or with decreased lifting or pushing power.

FLOW IMPEDANCE IN THE FILTER

If the filter is not chosen properly, both with regards to physical size and the characteristics of the

Figure 2-55 This piston shows evidence of scoring. This could be caused by hard particle contamination or possibly by hydraulic oil bypassing the seal and etching the piston as it traveled form one side of the piston to the other.

Figure 2-56 A wide variety of mobile equipment use fluid power systems to assist in the tasks assigned to the vehicle. Some vehicles, such as this lift truck, are merely mobile platforms for hydraulic systems. Other vehicles such as snow plows, dump trucks, and farm equipment use hydraulic systems to enhance their ability to perform their work.

media itself, then the filter can impede flow. Although great care is usually taken by the system designer to make sure the right filter was selected, parts availability or carelessness can often result in the wrong filter being installed in the system. Just using the wrong filter can have a huge impact on the flow of the hydraulic fluid.

FLUID VELOCITY CHANGES IN THE FILTER

Velocity is a function of flow rate and the diameter of the hose or line. The fluid is returning to the reservoir by means of one or more relatively small lines. When the fluid enters the filter, the relatively large volumetric capacity of the filter causes a decrease in the velocity of the fluid. This enhances both the filtration ability of the filter and its ability to cool the hydraulic fluid.

FLUID DAMAGE IN THE FILTER

The filter will never have a negative impact on the nature of the fluid, short of contaminating it when the filter is damaged.

FLUID CONTAMINATION IN THE FILTER

To say that the filter can contaminate the hydraulic fluid almost seems like an oxymoron. The basic task of the filter is to remove contaminates. It is the nature of most filters to become restricted and inhibit flow as it removes sediment and particles from the fluid. The collected material itself eventually begins to impede the flow of the fluid. At some point the restricted fluid flow will cause the pressure differential across the filter to become high enough to damage the filter, potentially ripping it apart and allowing shreds and shards of filter material to pass into the reservoir. Once these contaminates are in the reservoir they can be spread throughout the system by the pump (see RS 1-8, p. 42).

Summary

This has been a quick tour though a simple hydraulic system **(Figure 2-56)**. The intent of this chapter was to familiarize the reader with the most basic hydraulic components, how those components interact, and how the fluid behaves in the system to allow the system to accomplish its job. Subsequent chapters will have a detailed, deeper, and broader

discussion of each component and group of components. There will also be an extensive discussion and description of hoses, fittings, tools, and troubleshooting. This chapter was intended to make the interactions between components understandable. Understanding how a system works is fundamental to diagnosing a system when it fails to operate properly.

REVIEW SEQUENCE 1 · A Simple System: Component by Component

RS 1-1 The Reservoir The Journey of the fluid through the system begins in the reservoir. The reservoir has several purposes; chiefly, it is a storage vessel for system fluid. Usually made of metal or plastic, the reservoir has a relatively large volume that allows the oil passing through the reservoir, and that which is being stored in the reservoir, to have a very low velocity and thus allow for cooling, as well as permit sediment to settle out of the oil. The reservoir is sometimes pressurized with a gas to a point higher than atmospheric pressure to keep dust, biological debris, and other abrasive material out of the hydraulic system. Far more common is the reservoir that is simply exposed to atmospheric pressure. The pressure of the atmosphere applies a force to the surface of the fluid that helps move the fluid from the reservoir toward the next component in the system, the pump.

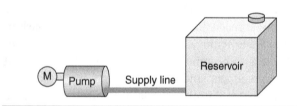

RS 1-2 The Supply Line and the Pump From the reservoir, atmospheric pressure along with the slight pressure decrease at the inlet of the pump causes the fluid to flow from the reservoir to the pump. The pressure in the reservoir is absolutely essential to make this flow happen. The supply line even in a system with a pressurized reservoir carries the fluid under very little pressure. The hose is usually made of rubber or plastic with little or no reinforcement. Once the fluid arrives at the pump, it's the job of the pump to move the fluid through the rest of the system. There is a common misconception among technicians, even highly experienced technicians, that pumps pressurize the fluid. All the pump does is move the fluid. When the moving fluid comes up against a restriction, the movement of the fluid stops—meanwhile, the pump keeps pushing. The pump attempting to push the fluid against an immovable object causes pressure to build up in the fluid.

At some point, the pump will begin to bypass internally—that is to say: At some pressure, it is easier for the fluid to slip past the gears, pistons, or vanes in the pump then it is to push the fluid through the restriction. For troubleshooting purposes, always remember that low pressure is rarely the fault of the pump. Low pressure is in almost all cases caused by a lack of restriction to flow or a failure of the pump to move the fluid. The small circle with an "M" in the middle represents an electric motor that drives the pump. The two most common power sources for a hydraulic pump are either an electric motor or an internal combustion engine.

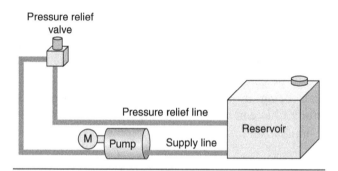

RS 1-3 Pressure Relief Valve In the event that the flowing fluid should come up against a significant restriction, the pressure can easily build between the pump and the area of restriction to the point where damage can occur to the pump, or the lines carrying the fluid can rupture, possibly damaging the system and causing injury. The pressure relief valve typically contains a spring-loaded pop valve that will open at a predetermined pressure. Once the pop valve opens, there is an alternate passageway back to the low pressure area that is the reservoir. This alternate passageway allows the pressure in the main line from the pump to the working part of the hydraulic system to drop significantly. Pressure relief valves always do the same job: They relieve pressure between the pump and the valves that control the direction of flow through the system. In systems known as closed center systems, when the directional control valve is in a position where fluid is now being directed to a piston or motor, all flow from the pump is blocked and the pressure relief valve must open in order to prevent overpressurization of the system. In all systems, including the closed center systems, the pressure relief valve serves to protect the system from overpressurization should the actuating device become stalled or a line become kinked.

REVIEW SEQUENCE 1

A Simple System: Component by Component (Continued)

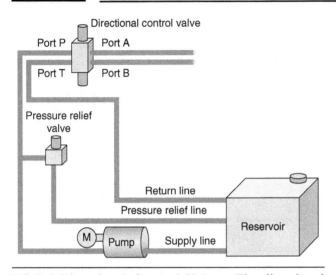

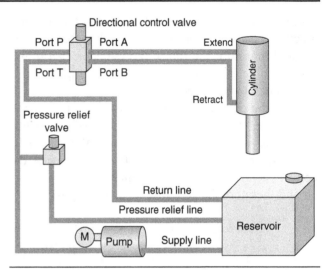

RS 1-4 Directional Control Valve   The directional control valve is one of the most common valves to control the operation of a hydraulic system's actuators. The flow from the pump enters the directional control valve typically at port P. For discussion purposes, assume initially that there is no fluid flowing to the actuator—no fluid flowing to port A or port B. If the valve is a closed center valve, the pressure will build up in the line from the pump to port P and eventually the pressure relief valve will open to direct fluid back to the reservoir to prevent overpressurization. If the valve is an open center valve, the fluid will simply enter at port P, pass through the directional control valve, and exit at port T. One way to remember the typical function of these two ports is that Port P is connected to the *p*ump and Port T is connected to the *t*ank.

When the directional control valve is moved from the idle (center) position to an operating position, such as straight through flow, then fluid can flow from port P to port A. After passing through the actuator, the fluid can then return to the reservoir by way of port B, through port T, and back to the reservoir. When the operator desires to reverse the direction of operation for the actuator, the directional control valve is put into the cross flow position. At this point, the fluid flows from the pump through port P, then to port B, and on to the actuator. After passing through the actuator, the fluid returns through port A, and crosses over to port T in returning to the reservoir. Although many combinations of directional control valves exist, the straight flow/cross flow directional control valve is the most common.

RS 1-5 Cylinder   Two basic actuators are used by fluid power systems. These are the cylinder and the hydraulic motor. A cylinder is typically used along with a lever or platform to push or pull a load. Essentially a hydraulic cylinder is a hollow tube, usually made of steel, with a snug fitting piston that slides back and forth within it. Hydraulic oil pressure and flow act upon the piston to move it back and forth within the tube. Connected to the piston is a rod that in turn will be connected to a lever or platform.

In this example, when the directional control valve is in the straight flow position, fluid will flow through port P of the directional control valve, exiting the valve at port A, and continuing to the extend side of the cylinder. The piston will then move, extending the rod outward. The fluid that was occupying the cylinder on the retract side of the piston is now forced back through port B, and then through port T of the directional control valve and back to the reservoir.

When the directional control valve is placed in the cross flow position, the fluid will enter through port P, and then pass over to port B and enter the retract side of the cylinder, forcing the piston back into the cylinder. The fluid that was in the cylinder on the extend side of the piston is then forced through port A. It then crosses over to port T and returns to the reservoir.

When the directional control valve is in the closed center position, one of two things will happen depending on whether the valve is open center or closed center. If the valve is an open center valve and is shuttled to the center position, oil in the retract and extend sides of the cylinder will be able to move freely back and

A Simple System: Component by Component (Continued)

forth between the two sides of the piston through the directional control valve. The end result is that the piston, and therefore the rod of the cylinder, can move when a force is applied to the rod. The piston and rod do not "lock" in position.

When the directional control valve is of the closed center type, the fluids on the two sides of the piston cannot move freely from one side of the piston to the other. The end result is when the closed center directional control valve is in the center position, the piston and therefore the rod are "locked" in position.

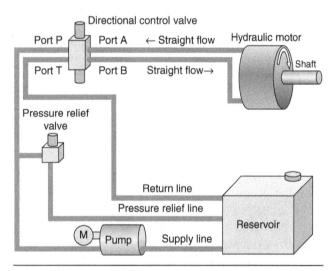

RS 1-6 Hydraulic Motor—Straight Flow Hydraulic motors are used to rotate hubs, wheels, and pulleys. When the directional control valve is in the straight flow position, oil will flow from port P through port A and on to the motor. After passing through the motor, the fluid will pass through port B, out port T, and return to the reservoir. As the fluid flows through the motor, the shaft on the motor will turn in a specific direction.

If the directional control valve is closed center, then when it is placed in the center position oil cannot flow through it. Fluid cannot flow through the motor, and the motor cannot push fluid through the directional control valve. The output shaft of the motor is therefore effectively "locked" in position. If the directional control valve is open center, then a torsional load on the output shaft of the motor can rotate the shaft and thus turn the motor into a pump. The open center valve will not block the flow of this fluid and the motor will be free to rotate.

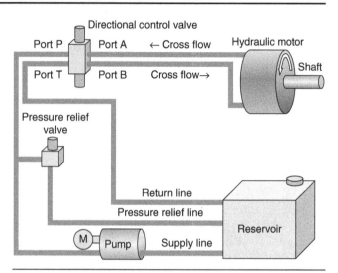

RS 1-7 Hydraulic Motor—Cross Flow When the directional control valve is moved to the cross flow position, oil will flow from port P to port B, through the motor and back to the directional control valve at port A, through the valve to port T, and back to the tank. This reverses the rotational direction of the motor's output shaft.

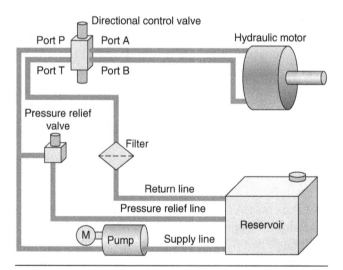

RS 1-8 Filter The hydraulic fluid filter is typically located on the return line of the hydraulic system. This makes sure that impurities are removed from the fluid before returning to the reservoir while not impeding the ability of the pump to draw fluid from the reservoir. Even though the filter is located on the return line, it can still have a significant impact on the fluid's ability to fluid.

Review Questions

1. Resistance to flow caused by a restriction in a hydraulic system causes:

   A. A drop in pressure.

   B. A rise in pressure.

   C. A drop in pressure before the restriction and a rise in pressure after.

   D. A rise in pressure before the restriction and a drop in pressure after.

2. Fluid moves from the reservoir to the pump because the pump creates a very large vacuum at its inlet port.

   A. True

   B. False

3. Which of the following items is most likely to cause damage to the hydraulic fluid?

   A. Heat

   B. Velocity

   C. Low pressure in the system

   D. Cold

4. Technician A says that low pressure in a hydraulic system will cause the system to be "weak." Technician B says that a restriction causing a low flow rate for the fluid can make the actuators of the system operate "slowly." Who is correct?

   A. Technician A only

   B. Technician B only

   C. Both Technician A and Technician B

   D. Neither Technician A nor Technician B

5. Force is required to move an object. In a hydraulic system, which measurements or characteristics determine force?

   A. Pressure

   B. Piston area

   C. Flow rate

   D. Size of the pump

6. Technician A says pressure throughout a fluid power system that contains a single flow path is constant. Technician B says the flow rate throughout a fluid power system that contains a single flow path is constant. Who is correct?

   A. Technician A only

   B. Technician B only

   C. Both Technician A and Technician B

   D. Neither Technician A nor Technician B

7. If the flow rate (volume per unit of time) remains constant when hydraulic fluid is forced through a reduced size hose or passageway, the velocity of the fluid in the reduced size hose or passageway must increase.

   A. True

   B. False

8. When a high pressure is present at the outlet port of the directional control valve, it is an indication that:

   A. The system is working normally.

   B. There is a restriction in the return line.

   C. There is excessive fluid in the system.

   D. The piston in the cylinder is too small.

9. Technician A says that more force can be obtained when a larger actuating piston is used. Technician B says that increasing the speed of rotation on the pump can increase the flow rate of the fluid. Who is correct?

   A. Technician A only

   B. Technician B only

   C. Both Technician A and Technician B

   D. Neither Technician A nor Technician B

CHAPTER

3 The Math

Learning Objectives

Upon completion and review of this chapter, the student should be able to:

- Calculate the surface area of a piston.
- Calculate the force or thrust of a piston at a given pressure.
- Calculate the volume of a cylinder.
- Calculate the travel speed of a hydraulic cylinder piston.
- Calculate the velocity of oil flow in a pipe.
- Calculate horsepower in a fluid power system.
- Calculate torque in a fluid power system.
- Perform calculations using the metric system.

Cautions for This Chapter

- Taking pressure and flow readings necessary to perform many of the calculations discussed in this chapter requires removing pressure tap plugs or hoses and lines. Before removing any plug, hose, or line, make certain there is no pressure behind the plug or in the line being disconnected.

- Always use high-quality purpose-built test equipment for taking measurements. "Homemade" or cobbled-together test equipment can not only be inaccurate, it can be dangerous.

- When inspecting a system, remember that the hydraulic fluid may be under high pressure.

- When working around a system with leaks, remember that hydraulic oil on the floor can be extremely slippery.

- Remember that when hydraulic fluid is under high pressure and when escaping under high pressure through a small leak, that leak may be invisible.

- Note that when hydraulic fluid is under high pressure and escaping through a small leak, the escaping high-pressure fluid can penetrate the skin and is toxic.

- Remember that hydraulic systems are designed to move large and heavy loads. These loads can be dangerous should they move or shift unexpectedly.

- ALWAYS wear safety glasses when working on fluid power systems.

Key Terms

area	horsepower	velocity
flow rate	torque	volume
force or thrust	travel speed	

INTRODUCTION

Although math is not required for most trouble-shooting procedures, the principles of how fluid power systems operate are not only linked to mathematical concepts, but are themselves mathematical concepts. There is evidence that fluid power systems date back thousands of years. The archeological record shows the fundamental concepts in use to supply water to many ancient cities. The first solid record we have of fluid power in use was in the work of Archimedes, a Greek born in Sicily in 287 BCE. First and foremost, Archimedes was a mathematician, and as such he was a man determined to explain his world in terms of equations and logic. He invented the "Archimedes screw," a crude but effective hydraulic pump, and was quoted as having said, "Give me a lever and I can move the world." What he meant is that with a long enough lever, any mass could be moved by anyone. The principle of the lever and the principle of hydraulic PAF (Pressure, Area, and Force) are very similar to one another. Both deal with using geometrical proportions to amplify force. However, modern hydraulic systems use levers to transfer force from the hydraulic cylinder (and sometimes a hydraulic motor) to the work.

The typical technician may never need to grab a calculator and work though the formulas covered in this chapter. But troubleshooting often demands an understanding of the engineering put into the design of the system being diagnosed. This understanding of the engineering allows the service technician to predict how the system should behave, how fast the piston and cylinder should move, how fast the motor should spin, and how much of a load it should be able to move.

CIRCLE FORMULAS

The circle plays heavily into the math and operation of fluid power systems. The force with which a hydraulic actuator can operate is determined by the **area** of the surface of the piston and the pressure acting on it. More pressure, more force. More area on the surface of the piston, more force.

The mathematical constant pi (π) dates from at least the nineteenth century BCE. As mentioned previously, it is the ratio of a circle's circumference to its diameter. In technologies like fluid power, this value is typically rounded off to 3.14, but in precision systems the value's exactness is increased by many decimal places. Pi is an interesting constant because supercomputers have been tasked with calculating pi to its final decimal place, or at least calculating it to the point where a repeating pattern is found. In calculations of up to one trillion decimal places; however, no repeating value has been found.

The area of a circle is calculated with the famous formula $\pi r^2 = A$ **(Figure 3-1)**. To make this easy, let us say that we have a piston with a diameter of 6 inches. The radius of the circle formed by the face of the piston is half of the diameter, or 3 inches. Squaring the radius is simply a matter of multiplying it by itself.

$$3 \times 3 = 9$$

Now multiply this square of the radius by pi, the 3.14 constant.

$$9 \times 3.14 = 28.26 \text{ square inches}$$

You may ask "Why is this important?" For every pound of pressure per square inch (psi) a piston with 28.26 square inches will be able to apply a force of 28.26 pounds. Taking this to an extreme, let us say that we place a 2-ton automobile on a piston 20 feet in

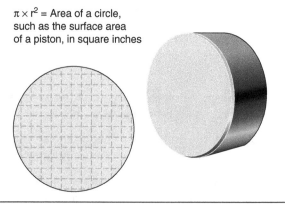

$\pi \times r^2$ = Area of a circle, such as the surface area of a piston, in square inches

Figure 3-1 The area of a circle is calculated by multiplying the radius (the distance from the center of a circle to its circumference) by itself, and then multiplying that result by the constant pi. Pi is equal to 3.14159.

diameter. In inches, the diameter would be 240 inches. The 120-inch radius squared is 14,400. Multiply that times pi and the result is 45,216 square inches. Less than one tenth of a psi would be required to lift the automobile. The catch is that the amount of fluid required to lift the automobile 1 inch would be enormous and unless the supply pump had a huge **flow rate**, the time to make the lift would be extremely long. When designing a system, the engineer must balance pressures, heat, flow rates, and the length of time required for the system to do its job. When troubleshooting a system, the technician must know how the engineer balanced these variables to understand how the system was intended to work. Piston area determines how much pressure will be required for the system to do its job.

The Rod Side Effective Area

In most fluid power cylinders, each piston has two sides **(Figure 3-2)**. The side that fluid applies force against to extend the actuating rod is called the *blind* side of the piston. Fluid pressure is applied to the rod side of the piston to retract the actuating rod. The ability of the piston to retract against a resistance is less than the ability to extend against a resistance because the surface area of the piston is less. The surface area of the rod side of the piston is reduced by the amount of area taken up by the rod's attachment to the piston.

To calculate the effective area of the rod side of the piston, it is necessary to first calculate the area of the

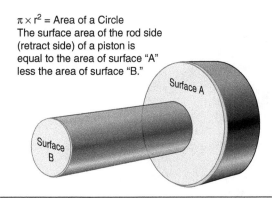

$\pi \times r^2$ = Area of a Circle
The surface area of the rod side (retract side) of a piston is equal to the area of surface "A" less the area of surface "B."

Surface A

Surface B

Figure 3-2 A fluid power system has two sides. On one side, the extend side of the piston, the full surface area of the piston is in contact with the pressure applied by the fluid. The rod side the surface area occupied by the rod is not exposed to the pressure provided by the fluid. As a result, the surface area of the piston is reduced by the area occupied by the rod. As a result, at a given pressure, the amount of force that can act on a load is decreased.

blind side of the piston, then subtract the cross-sectional area of the rod. For instance, in the previous example the piston had a diameter of 6 inches. The surface area was then calculated to be 28.26 square inches.

If the rod shaft was 3 inches in diameter, the radius of the rod would be 1.5 inches. Pi times the square of the radius would produce a result of $1.5 \times 1.5 \times 3.14$ or 7.065 square inches. Subtract the 7.065 inches from 28.26 square inches of piston area and the result will be 21.195 square inches of effective surface area on the retract side of the piston.

The Force or Thrust of Any Cylinder

Lifting automobiles with tiny pressures is not the goal of a modern hydraulic system. Yet in the previous example it can be seen how force is a function of pressure and area. The force (or thrust) formula is pressure times the area **(Figure 3-3)**. In the 28.26-square-inch piston area used in one of the preceding examples, the amount of force that would be applied with 1,000 psi of pressure on the piston is 28,260 pounds. This means that the cylinder with the 1,000 psi behind it could lift a dead weight in excess of 14 tons.

If the piston had a diameter of only 4 inches, as opposed to 6, the surface area would be 12.56 square inches. The 1,000 psi could now lift only 12,560 pounds, which is less than half the weight lifted by the 6-inch piston.

If the piston were 3 inches in diameter (half of the 6-inch piston), the area of the piston would be $1.5 \times 1.5 \times 3.14$ or roughly 7.1 square inches. At 1,000 psi, this cylinder would only be able to lift 7,100 pounds. In the case of the lifting capacity of a

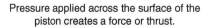
Pressure applied across the surface of the piston creates a force or thrust.

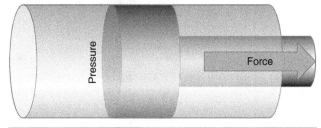
Pressure

Force

Figure 3-3 The amount of **force or thrust** that can be developed by a hydraulic cylinder is a function of the surface area of the piston and the amount of pressure applied to that surface area. If the surface area of the piston is 5 square inches and the applied pressure is 1,000 psi, then the amount of force that would be applied through the piston would be 5,000 pounds.

6-inch piston versus a 3-inch piston, the lifting capacity of the smaller is slightly over one fourth the lifting capacity of the larger. This is not a rule, however. For instance:

8-inch piston at 1,000 psi = 4 × 4 × 3.14

= 50,240 pounds of force

4-inch piston at 1,000 psi = 2 × 2 × 3.14

= 12,560 pounds of force

Even though the diameter of the 4-inch piston is one half the diameter of the 8-inch piston, just like the comparison of the 6-inch and the 3-inch, this time the lifting force is less than one fourth.

The Force or Thrust on the Retract Side

A 6-inch piston can produce 28,260 pounds of force at 1,000 psi to extend the actuator rod. If the rod is 3 inches in diameter, that leaves an effective area on the back side of the piston of 21.195 square inches. This means that the force that can be generated to retract will only be 21,195 pounds when the pressure is 1,000 psi **(Figure 3-4)**.

$$F = A \times PSI$$

F is the force or thrust, in pounds.

A is the piston net area, in square inches.

PSI is the gauge pressure.

Pressure is not applied to the area on the rod side of the piston. Therefore the force to the rod will always be lower than the force available to extend the rod at a given pressure.

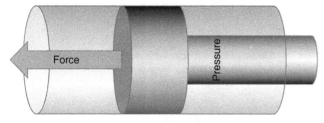

Figure 3-4 Since the rod of an actuating cylinder takes up space on the retract surface of the piston, the surface area of the retract side is always less than the surface area of the extend side. As a result, the maximum force that can be delivered to retract the cylinder is reduced as a function of the percentage of area taken up by the rod on the surface of the piston. If the piston has a surface area of 5 square inches and the rod takes up 40 percent of that surface area, then the potential of 5,000 pounds of force that exists on the extend side is reduced to only 60 percent of the 5,000 pounds. The force to retract this cylinder is only 3,000 pounds.

Cylinder Volume

The movement of a piston in a cylinder occurs when a volume of hydraulic fluid is moved into the cylinder against the piston. The volume required to move the piston an inch depends on the size of the piston. The larger the piston the more square inches of surface area there is and therefore the more volume there is in the cylinder. The greater the volume of the cylinder the greater the fluid flow rate to move the piston a given distance in a specified time. Simply put, the speed at which the cylinder will extend and retract is a function of flow rate and cylinder volume **(Figure 3-5)**.

Volume is measured in cubic length measurements such as cubic inches or cubic centimeters and is also measured in volumetric measurements like liters and gallons. The initial volumetric measurement will always be in cubic length measurements because we can easily measure these dimensions.

If the diameter of the piston were 4 inches, the area of the piston would be 12.56 square inches. If the length of travel for the piston were 20 inches inside the cylinder, then the volume of the cylinder at full extension would be 12.56 square inches times 20 inches

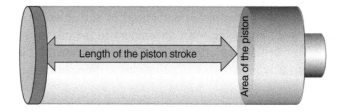

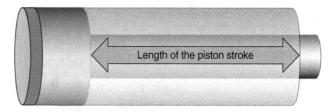

Figure 3-5 The volume of a cylinder is equal to the surface area of the piston times the length of the piston's stroke. Therefore, if the surface area of the piston is 5 square inches and the length of the stroke is 10 inches, the volume of the cylinder is 50 cubic inches. To calculate the volume of the rod side of the cylinder, it is necessary to calculate the volume of the rod itself. If the diameter of the rod is 2 inches, then the radius of the rod is 1 inch: 1 inch times 1 inch times 3.14159 square inches. The volume of the rod is therefore 31.4159 cubic inches. The 50 cubic inches of the extend side of the piston, less the 31.4159 cubic inches of the rod leaves a retract side volume of 18.5841 cubic inches.

or 251.2 cubic inches. Since there are 231 cubic inches in one U.S. gallon, the volume is calculated by dividing 251.2 by 231. The result is 1.087 gallons. Logically, if the flow rate of fluid into the cylinder were 1 gallon per minute, it would take slightly over 1 minute to extend the cylinder.

If the 4-inch piston had a 2-inch diameter rod, then the volume of the cylinder when it is retracting would be decreased by the volume taken up by the rod. A 2-inch diameter rod would have a cross-section surface area of 3.14 square inches. This is calculated, of course, by multiplying the radius of 1 inch by itself and then by 3.14. Multiply this cross-section surface area by the 20-inch length and the result is 62.8 cubic inches. The volume of the retract side is therefore the same 251.2 cubic inches less the 62.8 cubic inches occupied by the rod. The volume of the rod side of the cylinder is therefore 188.4 cubic inches. Since the volume of the rod side of the cylinder is 75 percent of the volume of the blind side of the cylinder, the piston and therefore the rod will move 25 percent faster on retract than it does on extend.

$$V = A \times L$$

V is the volume of the cylinder.

A is the area of the piston.

L is the length or distance that the piston can move in the cylinder.

Hydraulic Cylinder Piston Travel Speed

A major consideration in troubleshooting a system is to know how fast the piston moves in a hydraulic cylinder and therefore how fast the rod will move. Once the area of the piston and the flow rate **(Figure 3-6)**

The speed of the piston is equal to the flow rate of the fluid in cubic inches per minute divided by the area of the piston.

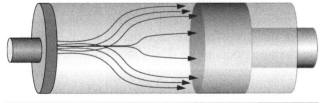

Figure 3-6 Speed of the fluid flow is a linear measurement. The flow rate of the fluid in a system is a cubic measurement such as gallons or cubic inches over a period of time, usually minutes. The speed or velocity of the fluid is its linear measurement over a period of time. To calculate the velocity of the fluid, simply divide the flow rate by the area of the piston. The result is the velocity.

of the system are known, then the speed of the piston can be easily calculated. If the system has a pump flowing 1 gallon per minute (231 cubic inches per minute) and the piston is a 4-inch piston with a 12.56-inch surface area then the speed of the piston is $231 \div 12.56$, or 18.39 inches per minute.

Again, because the rod takes up volume in the retract side of the cylinder, our 2-inch rod will net a piston surface area of 12.56 square inches less 3.14 inches. This is an area of 9.42 square inches. The piston speed when the rod is retracting would be 231 cubic inches divided by 12.56 square inches less the 3.14 square inches of the rod. This equals $231 \div 9.42$, or 24.5 inches per minute.

Noteworthy is the fact that the speed on retract is faster than the speed on extend. Not only is it faster because of the rod causing a reduced surface area of the piston on the retract side, but the reduced surface area also causes the available applied force to be reduced.

$$S = CIM \div A$$

S is the piston **travel speed**, in inches per minute.

CIM is the oil flow into cylinder, in cubic inches per minute.

A is the piston area in square inches.

The Velocity of Oil Flow in a Pipe

When the diameter of a pipe or other hydraulic fluid passageway changes, and the flow rate does not change, the **velocity** of the fluid must change **(Figure 3-7)**. For example, a system has a pump flowing 2 gallons per minute through a single path in the system. The hose through which the fluid passes has a cross-sectional area of 2 square inches. The velocity will equal

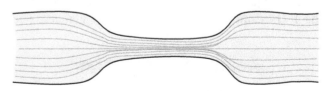

Figure 3-7 The flow rate of fluid is a constant through any given branch of a hydraulic system. When the pathway for the fluid narrows, the fluid must increase velocity to maintain the flow rate. When the pathway gets larger, the velocity of the fluid decreases. As a result, the velocity (or speed) of the fluid flowing through a given branch of the system will increase and decrease as the size of the pathway decreases and increases. Speed changes inversely as the hose or pathway size changes.

2 (gallons per minute) × 0.3208 (a constant). This equals 0.6416. Multiply the result by the area, 2 square inches, and the calculated flow rate is 1.2832 feet per second. This is the speed of the fluid flowing through the hose where the cross-sectional area is 2 square inches.

The inquisitive mind might wonder where the constant 0.3208 comes from. The constant is a mathematical shortcut. We are starting with a flow rate in gallons, minutes, and inches; we are ending up with an answer in feet per second. The 0.3208 is derived from 231, which is the number of cubic inches in a U.S. gallon; the conversion of feet to inches (12); and the conversion of inches to seconds (60).

231 cubic inches in a gallon ÷ (12 inches in a foot × 60 seconds in a minute) = 0.3208.

$$V = GPM \times 0.3208 \div A$$

V is the oil velocity in feet per second.

GPM is the flow in gallons per minute.

A is the inside area of the pipe in square inches.

Hydraulic (Fluid Power) Horsepower

Power is a measurement of force over a period of time. In the case of a fluid power system it is a measurement of flow rate and pressure over a period of time. Scientists have calculated that one **horsepower** equals 550 foot-pounds per second. Since measurements in the typical fluid power system are usually made in cubic inches and square inches instead of feet, and in minutes instead of seconds, some conversions are necessary.

Using the formula in **Figure 3-8**, flow rate of 4 gallons per minute with a pressure of 2,000 psi would yield 4.67 horsepower.

The inquisitive reader might wonder from where the 1,714 came. This is not an arbitrary number. The

$$Horsepower = \frac{Gallons\ per\ minute \times psi}{1,714}$$

Figure 3-8 The fundamental purpose of a fluid power system is to do work. The ability of a system to do work is measured in horsepower. In a fluid power cylinder system, horsepower can be calculated by multiplying the pressure being applied to the piston in the cylinder by the amount of pressure. The 1,714 found in the divisor is simply a conversion constant that is used to match up the various units of measure in this formula.

550 foot-pounds per second stated earlier is the measured definition of a horsepower. In fluid power systems, most measurements are done in inches and minutes. The 1,714 is called a constant and is used to make conversions from seconds to minutes and feet to inches without a lot of complex math.

$$1,714 = (550[\text{foot pound per second}]) /$$
$$((231[\text{gallons to cubic inches}]$$
$$\div (12[\text{feet to inches}] \times 60[\text{minutes to seconds}]))$$

One could add this element into their calculations each time they try to calculate horsepower, however the conversions of minutes to seconds, of feet to inches, and from gallons to cubic inches never changes. Therefore, using 1,714 is a handy and accurate shortcut.

$$HP = PSI \times GPM \div 1,714$$

PSI is the gauge pressure in pounds per square inch.

GPM is the oil flow in gallons per minute.

The Relationship between the Displacement and Torque of a Hydraulic Motor

Hydraulic motors are important components in modern hydraulically operated winches, turntables, and traction devices. The portion of a hydraulically operated system through which fluid flows is tied to the laws of physics associated with pressure, force, and area. The mechanical portion of the system follows the laws of physics associated with the lever and **torque**. When it comes to the hydraulic motor, the principle of the lever is a little less obvious **(Figure 3-9)**.

All hydraulic motors consist of a series of chambers connected around a center point. These chambers may be formed by the teeth of gears, vanes, or even rollers that serve as a surface that not only separates the chambers but one on which the hydraulic fluid and resulting force can act. As with a piston, the greater the surface area of these separating surfaces, the greater the force that will act on them. The larger the surface area of these separators, the larger the chambers will be that they separate. The larger the chambers, the greater the amount of fluid the chambers will hold. The amount of fluid that can be held by each chamber times the number of chambers carrying force-producing fluid is the displacement of the motor. Since the force

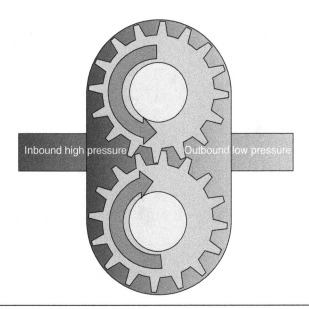

Figure 3-9 The pressure differential across a hydraulic motor is the difference between the pressure on the inbound side of the motor and the pressure on the outbound side of the motor.

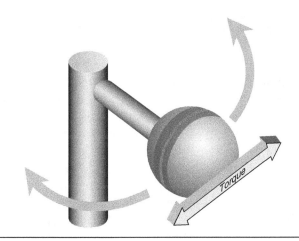

Figure 3-10 Torque is the arcing or circular force applied to, or resulting from, a rotating object, or an object that is traveling in an arc.

of leverage in a motor is rotational in nature, the measurement of this force is actually a measurement of torque **(Figure 3-10)**.

The formula for calculating the torque of a motor when the displacement is known is:

Torque =

$$\frac{\text{Displacement in cubic feet per revolution} \times \text{Pressure drop across the motor in psi}}{24\pi}$$

As with many formulas, the constant 24 is used as shorthand to convert measurement standards. The 24 is

actually 12 × 2. The 12 is used to convert the feet of the cubic feet into the inches of the pounds per square inch. The 2 in the denominator (dividing by 2) is used to compensate for the fact that only half of the total displacement of the motor is doing any work, the return side of the motor is simply pushing fluid back to the reservoir.

If a motor has a displacement of 30 cubic inches, the pressure into the motor is 1,100 psi and the pressure on the outbound side of the motor is 100 psi, which yields a pressure drop across the motor of 1,000 psi. The calculation is thus made as follows:

$$\text{Torque} = \frac{30 \times 1,000}{24\pi}$$

$$\text{Torque} = \frac{30,000}{75.36}$$

Torque = 398 pound feet of torque

$$\mathbf{T = D \times PSI \div 24\pi}$$

T is torque in foot pounds; D is displacement in cubic inches per revolution.

PSI is the pressure difference across the motor.

$$\pi = 3.14$$

Torque and Horsepower Relations at RPM

There would only be rare instances when a mobile equipment technician might need to calculate or even be concerned with the relationship between torque and horsepower. Torque is a measurement of how much power a motor, in this case a hydraulic motor, produces to move an object. Horsepower is a measurement of how fast that object can be moved.

The formula for describing the relationship between torque and horsepower is defined in **Figure 3-11**.

Like many formulas, this formula contains a constant. Again, this constant is used to convert units of measurement. One horsepower is defined as 550 foot pounds per second of torque. Torque is a measurement of rotating force applied over a distance. Horsepower is a measurement of linear force over a minute.

$$\text{Torque in ft. lbs.} = \frac{\text{Horsepower} \times 5,252}{\text{Motor RPM}}$$

Figure 3-11 In any mechanical system, horsepower can be calculated from a known torque value.

The constant converts rotational measurement to a linear measurement and the minutes of RPM into seconds. To convert the rotational measurement to linear measurement, the formula for finding the length of a circumference is used. This formula is radius times 2 times pi. We will use a radius of one since we are developing a constant.

$$\textbf{radius} \times \textbf{2} \times \pi = \textbf{1} \times \textbf{2} \times \textbf{3.14159}$$
$$= \textbf{2} \times \textbf{3.14159} = \textbf{6.28318}$$

The linear measurement is then divided by 60 to convert for minutes to seconds.

$$\frac{\textbf{6.28318}}{\textbf{60}} = \textbf{0.10471}$$

The linear measurement per second (0.10471) is then divided by foot pounds of torque to horsepower (550).

$$\frac{\textbf{550}}{\textbf{0.10471}} = \textbf{5,252}$$

The final constant of 5,252 is, of course, rounded off for convenience.

Using this formula let us determine how much torque a motor putting out 300 horsepower can produce at 2,000 rpm.

$$\textbf{Torque in foot pounds} = \frac{\textbf{30 horsepower} \times \textbf{5,252}}{\textbf{2,000 RPM}}$$

$$\textbf{Torque} = \frac{\textbf{157,560}}{\textbf{2,000}}$$

$$\textbf{Torque} = \textbf{78.78 foot pounds}$$

$$\textbf{T} = \textbf{HP} \times \textbf{5,252} \div \textbf{RPM}$$

$$\textbf{HP} = \textbf{T} \times \textbf{RPM} \div \textbf{5,252}$$

Torque values are in foot pounds.

METRIC FORMULAS

The metric system has evolved from a system of measurement first adopted in late eighteenth-century France. The English System, or as it is known outside of the United States, the Imperial System, is an extremely convoluted system derived from the many measurement systems that were brought to the British Isles by invading armies ranging from the Romans to the Norse to the Normans. The metric system was an effort to bring order and international agreement. In a little over 200 years this system of measurement has been officially adopted by every country in the world except three. The countries yet to make an official

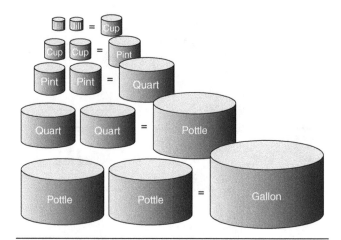

Figure 3-12 The "American" measuring system is based on the arbitrary and originally imprecise units of the British Imperial system.

conversion are Myanmar, sometimes referred to as Burma, Liberia in West Africa, and the United States.

Much of the English System was based on the size of body parts. For instance, the mouthful was once a common measurement of liquid volume. It was roughly equivalent to one fluid ounce. There were two mouthfuls in a jigger, two jiggers in a jackpot, two jackpots in a jill, two jills in a cup, two cups in a pint, two pints in a quart, two quarts in a pottle, and two pottles in a gallon. Therefore, a gallon is equal to 256 mouthfuls. To further complicate the issue, there is a British (Imperial) gallon and an American gallon **(Figure 3-12)**. The British gallon is equal to approximately 1.2 American gallons. This could be interpreted to mean that the British have bigger mouths than the Americans? In the English system, length measurements are based on the length of a barleycorn and increase in units of three, for the most part, instead of units of two like the liquid volume measurements.

Metric units of measurement are mostly based on scientific constants or universally consistent measurements. The length of an individual barleycorn can vary from plant to plant, field to field, region to region and season to season. The meter was originally based on a length that was equal to 1/40,000,000 the circumference of the earth at the equator. The gram is the weight of one hundredth of a meter square volume (cubic centimeter) of water. A thousand cubic centimeters is a liter. Most pressure measurements are based on the bar. Although the bar is not officially a measurement that is part of the metric system, it is an internationally accepted unit of measurement because it is almost exactly equal to the atmospheric pressure at average sea level. The Newton is a measurement of force. One Newton is the force that produces an acceleration of

Meter

decameter = 10 meters (a measure used in naval artillery)
hectometer = 100 meters (not a commonly used measure)
kilometer = 1,000 meters
decimeter = 1/10 of a meter
centimeter = 1/100 of a meter
millimeter = 1/1,000 of a meter

Liter

deciliter = 10 liters (not a commonly used measure)
hectoliter = 100 liters
kiloliter = 1,000 liters (not commonly used)
deciliter = 1/10 of a liter
centiliter = 1/100 of a liter
milliliter = 1/1,000 of a liter

Figure 3-13 Metric measurements are based on precise and incremental uniform standards. In spite of this most hydraulic system measurements and descriptions for North American machinery utilizes the "American" measurement system.

1 meter second per second on a mass of 1 kilogram. The metric equivalent of horsepower is the kilowatt. A watt is equal to 1 joule per second; a kilowatt is 1,000 watts. There are approximately 746 watts (0.746 kilowatts) in 1 horsepower.

Circle Formulas

When using metric measurements, the only thing that really changes is the units of measurement **(Figure 3-13)**. A piston with a diameter of 10 centimeters will have a radius of 5 centimeters. Pi times the square of the radius will yield the surface area of the piston. Using 3.14159 for increased accuracy, this should be multiplied by the square of the radius, which is 25. The result is 78.54 square centimeters. Multiply this times the length that a piston moves inside a cylinder and you get the swept volume. Seventy-five centimeters of movement (a little less than 30 inches) would render a volume of 5,890.5 cubic centimeters. This is equal to 5.8905 liters.

Force or Thrust of a Cylinder

Just like with the Imperial system of measurement, force is pressure multiplied by area.

Newtons of Force = Area in Square Centimeters × Pressure in Bar × 10

In the preceding formula, the 10 is a constant used to adjust units of measurement just as was done in the Imperial system. An area of 78.54 square centimeters times 100 bar times the conversion factor of 10 would yield a force of 78,540 Newtons. To compare this with the more familiar pounds of force measurement, 78,540 Newtons is roughly equivalent to 17,656 pounds of force.

Hydraulic Cylinder Piston Travel Speed

The speed of a hydraulic cylinder is dependant on the area of the piston in the cylinder and the flow rate in the system. For this calculation, the flow rate is measured in cubic decimeters per minute. A U.S. gallon is equal to 3.785412 cubic decimeters. A cubic decimeter is more commonly known as a liter.

Speed in meters per second
$$= \frac{\text{Volume in cubic decimeters per minute}}{\text{Piston area in square centimeters}}$$

Using the 10-centimeter cylinder mentioned earlier, the area of the piston is 78.54 square centimeters. If the flow rate is 15 liters per minute (15 cubic decimeters per minute), then the piston **travel speed** will be:

$$\frac{\textbf{15 cubic decimeters per minute}}{\textbf{78.54 square centimeters}}$$

$$= \textbf{0.1909 meters per second}$$

These values yield a piston speed of 0.1909 meters per second, or 19.09 centimeters per second.

Velocity of Oil Flow in a Pipe

The metric formula for calculating the velocity of oil flow in a pipe is:

Oil velocity in meters per second
$$= \frac{\text{oil flow in cubic decimeters per minute}}{6 \times \text{inside area pipe in square centimeters}}$$

The constant "6" in this formula is used to convert from minutes to seconds, and at the same time from decimeters to centimeters.

$$\frac{\textbf{15 cubic decimeters per minute}}{\textbf{6} \times \textbf{78.54 square centimeters}}$$

$$= \textbf{0.032 meters per second}$$

In centimeters per second, the velocity of the fluid is 3.2 centimeters per second. This is about 1.26 inches per second.

Hydraulic (Fluid Power) Kilowatts (Horsepower)

The metric equivalent to horsepower is kilowatts. One horsepower is roughly equivalent to 0.746 kilowatts.

Kilowatts =
Pressure in Bar × Flow in cubic decimeters per minute

The 600 is a constant used to convert time and other measurements.

$$\frac{100\ \text{Bar} \times 15\ \text{cubic decimeters per second}}{600}$$

$$= 2.5\ \text{kilowatts}$$

The horsepower equivalent of 2.5 kilowatts is about 3.35 horsepower.

Relationship between Displacement and Torque of a Hydraulic Motor

The first variable that needs to be measured in this calculation is pressure. This reading needs to be in Bar. If the reading is in kilopascals, for instance 20,000 kilopascals, simply divide by 100 to convert the reading to bar. In this example, this would yield 200 Bar. The displacement reading for the motor needs to be measured in cubic centimeters. The formula is then:

Newton meters

$$= \frac{\text{Pressure in Bar} \times \begin{array}{c}\text{Displacement volume}\\\text{in cubic centimeters}\end{array}}{20 \times \pi}$$

Therefore:

$$\frac{200\ \text{Bar} \times 40\ \text{cubic centimeters of displacement}}{20 \times 3.14159}$$

$$= \text{Torque in Newton meters}$$

$$\frac{200 \times 40}{62.8318} = \frac{8,000}{62.8318} = 127.37$$

The 20 is the conversion factor resulting from converting from pressure readings in Newton meters squared to bar, and from converting square centimeters to square meters so that all the measurements are based on the same units of measure. The fact that the motor only uses half of its total displacement to create power, or torque, is also factored into this constant.

Torque and Horsepower Relations at RPM

In the Imperial version of this formula, we used a constant of 5,252 to convert time, area, rotation, and

length measurements. The equivalent constant in metric is 9,543.

$$\text{Torque in Newton meters} = \frac{\text{kilowatts} \times 9,543}{\text{RPM}}$$

Therefore:

$$\frac{2.5\ \text{kilowatts} \times 9,543}{2,000\ \text{RPM}}$$

$$= 119.29\ \text{Newton} - \text{meters of torque}$$

RULES OF THUMB

Horsepower for Driving a Pump

For every 1 HP of drive, the equivalent of 1 GPM at 1,500 psi can be produced.

Horsepower for Idling a Pump

To idle a pump when it is unloaded will require about 5 percent of its full-rated horsepower.

Compressibility of Hydraulic Oil

Volume reduction is approximately 0.5 percent for every 1,000 psi of fluid pressure.

Compressibility of Water

Volume reduction is about 1/3 percent for every 1,000 psi pressure.

Wattage for Heating Hydraulic Oil

Each watt will raise the temperature of 1 gallon of oil by 1°F per hour.

Flow Velocity in Hydraulic Lines

Pump suction lines 2 to 4 feet per second

Pressure lines up to 500 psi, 10 to 15 feet per second

Pressure lines 500 to 3,000 psi, 15 to 20 feet per second

Pressure lines over 3,000 psi, 25 feet per second

All oil lines in air-over-oil system, 4 feet per second

Summary

Most mobile equipment technicians will work their entire careers without having to pick up a calculator and work through any of the math formulas discussed in this

chapter. Computer software, as well as many references, are available to perform the mathematics for the technician. However, an understanding of the math

TABLE 3-1: PUSHING AND PULLING FORCES FOR TYPICAL CYLINDERS

Cylinder area, square inches	0.754	3.142	7.065	12.57	19.64	28.27
Pushing force, Lbs @ 1,000 psi	780	3,142	7,065	12,750	19,640	28,270
Pulling force, Lbs @ 1,000 psi	343	2,034	5,951	11,460	18,530	27,160

TABLE 3-2: FORMULAS THAT MAY BE HANDY

To Solve For	Enter This into the Calculator
PTO output speed	Engine RPM × PTO% = PTO Speed
Engine input speed	Desired PTO speed ÷ PTO% = Required Engine Speed
Mechanical horsepower	T × RPM ÷ 5,252 = HP
Torque	HP × 5,252 ÷ RPM = T
Revolutions per minute	HP × 5,252 ÷ T = RPM
Capacity of reservoir	Li × Wi × Di ÷ 231 = G
Capacity of cylinder	πr^2 × Li ÷ 231 = G
Pump output horsepower	GPM × PSI ÷ 1,714 = HP
Pump input horsepower	GPM × PSI ÷ 1,714 ÷ E = HP

related to fluid power systems provides technicians with a better understanding of how they work. A better understanding inherently means better troubleshooting skills. Being familiar with the math makes it possible for the technician to estimate how the system should be able to operate. Documentation for most systems provides this information. However, the documentation is often incomplete or missing. A little knowledge of pump capacity, line sizes, cylinder size, and/or motor specifications along with a little math, even a rough estimation, can allow the technician to supply his or her own documentation (**Table 3-1** and **Table 3-2**).

Review Questions

1. The diameter of a cylinder's piston is 4 inches. What is the surface area of the piston?
   - A. Diameter times pi divided by 2 = 2.094 square inches
   - B. Diameter divided by 2 times diameter divided by 2 times pi = 12.566 square inches
   - C. Diameter times diameter divided by pi = 5.092 square inches
   - D. Pi squared times diameter = 39.478

2. The volume of a cylinder is:
   - A. Piston area times length.
   - B. Piston diameter times length.
   - C. Diameter times pi times length.
   - D. Same as the swept length.

3. The travel speed of a piston in a cylinder is primarily determined by:
   - A. Its length.
   - B. Its diameter.
   - C. Its directional control valve position.
   - D. The fluid flow rate.

4. In the formula for calculating hydraulic system horsepower, the number 1,714 is used. What does this number represent?

   A. This is the number of horsepower in a gallon of hydraulic fluid (oil).

   B. This number is a constant and is used to adjust for differences in measurement units.

   C. This number converts cubic inches into gallons.

   D. This number is the metric conversion factor.

5. The amount of force that can be generated in a fluid power cylinder is determined by:

   A. Pressure.

   B. The area of the piston.

   C. The pressure and the area of the piston.

   D. The flow rate.

6. Being familiar with metric measurements is important:

   A. Because the math is easier.

   B. Because the United States is only one of three countries in the world that does not use the metric system.

   C. Because foreign built machinery is often built to metric tolerances.

   D. All of the above

7. The velocity of fluid flowing through a pipe:

   A. Increases as the diameter of the pipe decreases.

   B. Decreases as the diameter of the pipe decreases.

   C. Increases as the diameter of the pipe increases.

   D. Decreases as the diameter of the pipe increases.

   E. A and D

   F. B and C

8. Torque is a measurement of:

   A. Rotating power.

   B. Tightness.

   C. Rotary movement.

   D. Linear movement around a pivot point.

9. Pi (π) is:

   A. An arbitrary number that seems to work well with fluid power formulas.

   B. The relationship between the radius of a circle and the diameter of a circle.

   C. The relationship between the diameter of a circle and its circumference.

   D. Exactly 3.14159.

10. PSI $\times$ GPM $\div$ 1,714 is the formula for calculating:

   A. RPM.

   B. Liters.

   C. Fluid velocity.

   D. Horsepower.

CHAPTER

ANSI Symbols

Learning Objectives

Upon completion and review of this chapter, the student should be able to:

- Read common hydraulic schematics and diagrams.
- Describe the fluid flow through a hydraulic circuit.
- Draw a simple hydraulic circuit.

Cautions for This Chapter

- Many symbols are similar. If in doubt about the meaning of a symbol, contact the publisher of the diagram.
- Always use the most recent version of the diagram for the equipment with which you are working. Remember to match the diagram to the model, production run, and production date of the equipment being diagnosed.
- Equipment is often modified in the field without informing the manufacturer or distributor of the equipment; be sure to inspect the equipment thoroughly and be sure the equipment matches the diagram.

Key Terms

accumulator

ANSI

bidirectional

pressure compensation

pressure reducing valve

relief valve

sequence valve

symbol

FLUID POWER GRAPHIC SYMBOLS AND COMPONENTS

Fluid power **symbols** are used to create detailed schematics of hydraulically operated systems. Although some companies choose to use proprietary symbols, most hydraulic systems manufacturers use standard **ANSI** (American National Standards Institute)

symbols. At first these symbols may seem somewhat arbitrary and cryptic. However, just like an alphabet, these symbols—even without supporting text—can explain exactly how a system is designed to function. This chapter will begin with each of the basic symbols, and then add the additional markings that represent increasingly specialized variations of each component or valve.

Components

PUMPS

The Basic Pump Symbol

The basic symbol for a pump is a circle with a single triangle indicating the direction of flow **(Figure 4-1)**. This symbol represents a *fixed displacement* **bidirectional** *pump*. With this type of pump, every rotation of the pump moves the same amount of fluid at all times. Increase the speed of the pump and the gallons per minute flow rate will always increase. Decrease the speed of the pump and the flow rate will always decrease. Additionally, the pump has been designed to flow fluid in only one direction.

The Bidirectional Pump

If a second triangle is added 180 degrees opposite of the first triangle and pointing outward, it indicates that the pump is a bidirectional pump **(Figure 4-2)**. A bidirectional pump will reverse the direction of flow when the rotational direction of the pump is reversed. These are typically used in small systems and are typically powered by a DC electric motor. When the current passing through the electric motor is reversed, the direction of the pump is reversed and therefore the direction of movement in a fluid power system is reversed. Use of this type of pump along with a reversing electric motor reduces the number of control valves the system needs.

Variable Displacement Unidirectional Pumps

In many systems there is a need to change the flow rate while the pump is engaged. This can be done by varying the RPM of a fixed displacement pump. In many applications, the power source—for instance, a

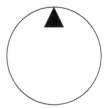

Figure 4-1 A fixed displacement unidirectional pump.

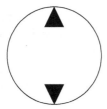

Figure 4-2 A fixed displacement bidirectional pump.

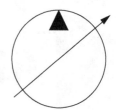

Figure 4-3 A variable displacement unidirectional pump.

diesel engine—may have other duties, such as driving a truck down the road, which may be a more critical determinant of speed than the hydraulic pump. For applications like these, variable displacement pumps are available **(Figure 4-3)**.

The arrow that bisects the pump symbol diagonally indicates the variable nature of the pump. Diagonal arrows drawn through basic symbols and extending beyond the limits of those symbols indicates that this symbol represents a component that has variable characteristics.

Variable displacement pumps have the ability to change their flow rate based on the desires of the operator or the demands of the system. A common place to find a variable displacement pump would be in a system where the speed of the hydraulic pump's power supply might change drastically while the pump is an operation, but the flow rate would need to remain the same.

Variable Displacement over Center Pumps

A variable displacement pump that is capable of flowing fluid in either of two directions is called a variable displacement over center pump **(Figure 4-4)**. Simply, a variable displacement over center pump is a variable displacement **bidirectional** pump.

Variable Displacement Pressure-Compensated Pumps

The vertical arrow within the pump symbol indicates that this pump is pressure compensated **(Figure 4-5)**. **Pressure compensation** means that the flow rate changes as the pressure changes. In a typical configuration, as the pressure increases in the system

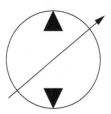

Figure 4-4 A variable displacement over center pump.

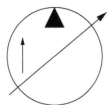

Figure 4-5 A variable displacement pressure-compensated pump.

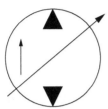

Figure 4-6 An over center with pressure-compensation pump.

downstream of the pump, the pump will adjust for the increasing pressure by decreasing the flow rate. This also means that when the pressure decreases in the system, the pump will automatically respond by increasing the flow rate. The end result of this action is that the system will maintain a relatively constant force even as conditions cause the flow rate to change.

Over Center with Pressure-Compensation Pumps

The over center with pressure-compensation pump **(Figure 4-6)** combines the characteristics of pressure compensation and reversibility into one pump. The end result is a pump that has the ability to reverse direction, thereby reversing the direction of flow while also being able to alter the flow rate as the pressure changes in the system.

CYLINDERS

Hydraulic cylinders are the most common of the fluid power operating components. When the pressure of the hydraulic fluid acts across the surface area of the piston inside the cylinder, the forces generated move the piston and the rod connected to the piston to perform the work. Most cylinders are double acting, that is to say that fluid pressure is applied to one side of the piston to extend the rod, and fluid pressure is applied to the other side of the piston to retract the rod. Single acting pistons will always require a mechanical force such as the force of the load, the force the rod is acting on, or air pressure to return the rod and the piston to its rest position.

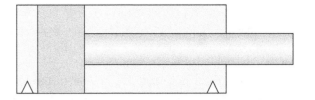

Figure 4-7 A double acting cylinder.

Double Acting Cylinders

The double acting cylinder **(Figure 4-7)** is probably the most common type of cylinder in use on modern hydraulic equipment. In a double acting cylinder, hydraulic fluid pressure and flow are required to extend the rod of the cylinder, and hydraulic pressure and flow are required to retract the rod of the cylinder. Because the route takes up surface area on the retract side of the piston and because it also takes up volume inside the cylinder at the given flow rate, the piston rod will always take longer to extend than it does to retract. In most system designs, this is acceptable because usually the heaviest load and the hardest work is being done while the piston and rod are extending. The greater surface area of the extend side piston provides a greater working force than does the retract side of the piston.

Double Acting Cylinders with Double End Rods

The double acting cylinder with a double end rod **(Figure 4-8)** is used in applications where two devices must be moved simultaneously but in opposite directions. It also can be used with one end not attached to anything so that the speed and force of both the retract and the extend functions are identical.

Single Acting Cylinders

Single acting cylinders **(Figure 4-9)** are typically used where the weight of the load being lifted by the

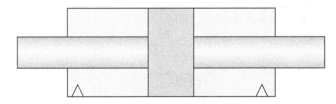

Figure 4-8 A double acting cylinder with double end rod.

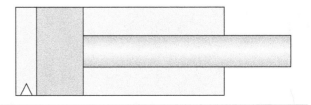

Figure 4-9 A single acting cylinder.

cylinder on the extend stroke will cause the piston rod to move back to the retracted position because of gravity. Retraction would occur when the fluid on the extend side of the cylinders vented back to the reservoir. When the fluid is vented, the weight of the load, or the weight of the arm, or the weight of the lever being actuated by the cylinder will force the rod and the piston back to its rest position.

Double Acting Cylinders with Cushions

Cushions are sometimes incorporated into both single and double acting cylinders **(Figures 4-10 and 4-11)**. The job of the cushion is to slow the movement of the piston as it approaches the full extend position or the full retract position, so as to soften the stopping action of the piston and rod. This prevents the piston from damaging the cylinder as it reaches the end of its movement. This cushion can be a physical rubber (or similar material) damper, or it can be done with orifices.

If orifices are used to dampen the movement near the end of the piston travel, they can progressively reduce the flow rate. The orifices for reducing the extend function will be found on the retract, or rod, side of the piston. There will be at least two orifices, with one arranged closer to the end of the cylinder than the other. These are the orifices through which the fluid in the retract side of the piston returns to the reservoir. As the piston nears the end of its travel, it covers the larger of the two orifices and prevents fluid from flowing through it. Since fluid can only flow into

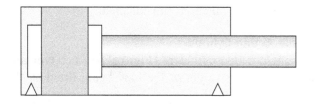

Figure 4-10 A double acting cylinder with cushion.

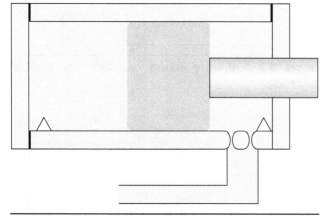

Figure 4-11 A double acting cylinder with cushion.

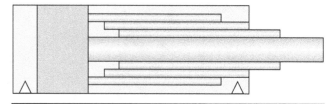

Figure 4-12 A telescoping cylinder.

the extend side of the cylinder as fast as fluid can escape from the retract side of the piston, the piston must slow. This buffers the stopping of the piston by slowing the movement before it reaches the end of its travel.

Telescoping Cylinders

Telescoping cylinders **(Figure 4-12)** are used when full extension of the cylinder needs to be more than twice the length of the cylinder. Despite telescoping cylinders being inherently weaker than a one-piece rod cylinder, they are frequently used in some extremely heavy duty applications, such as truck-mounted cranes and refuse truck compactors. The rod of a telescoping cylinder is made up of tubular pieces, each progressively smaller than the next, with the final rod connected to the actuated device nested in the center with the next smaller tube, which is nested in the next smaller tube, which is nested in the next smaller tube, and so on.

Inherently, the piston surfaces of the retract end of a telescoping cylinder are very small. This means there must be little resistance during retraction or the pressure in the system will have to increase. It is very common on telescoping cylinder applications for the pressure to rise significantly during retraction.

MOTORS

Fixed Displacement Unidirectional Motors

Hydraulic motors are common devices used in turntables, winches, and other systems that involve rotation. They are also commonly used on rope or cable reels.

Just like pumps, many types of hydraulic motors exist. The simplest type of hydraulic motor is known as the fix displacement unidirectional motor **(Figure 4-13)**.

Figure 4-13 A fixed displacement unidirectional motor.

Figure 4-14 A fixed displacement bidirectional motor.

A fixed displacement hydraulic motor will always yield the same amount of power and speed at any given flow rate and pressure differential. A unidirectional hydraulic motor, just like the unidirectional hydraulic pump, is designed to rotate in only one direction. Since one of the primary advantages of the hydraulic motor is its ability to rotate a device or component first in one direction and then in the other, unidirectional motors are not very common.

Fixed Displacement Bidirectional Motors

Because of its ability to rotate the component or device in one direction and then the other, the fixed displacement bidirectional motor **(Figure 4-14)** is the most common of all hydraulic motors. They are typically used to rotate and operate winches. Just like two fixed displacement unidirectional motors, this motor will always yield the same amount of power and speed at any given flow rate and any given pressure differential.

When the hydraulic directional control valve is moved to a position to reverse the flow, the direction of the bidirectional motor will reverse. This is handy, and even necessary, on a winch, and is necessary to both reel out and reel in the cable or rope. Bidirectional motors are also used in applications such as tank turrets and ladders on fire trucks.

Variable Displacement Unidirectional Motors

The variable displacement unidirectional motor **(Figure 4-15)** is a lot like a variable displacement pump internally. Often these motors are piston motors with the pistons riding on a variable angle swash plate.

Figure 4-15 A variable displacement unidirectional motor.

Figure 4-16 A bidirectional motor with pressure compensation.

As the angle on the swash plate changes, the amount that each system can move through a rotation also changes. Therefore, the flow of the fluid through the system will move each piston a little less, thus reducing the volume required to make a full travel movement on the system, allowing a greater RPM for the motor at the given flow rate.

Variable displacement motors can be used very effectively in applications where motor speeds are required to go from zero to very high. When the swash plate is at the minimum angle, the pistons barely displace, therefore a small amount of flow will cause the motor to spin at a high speed. Although the motor is turning at a high speed with the swash plate at its minimum angle, the fact there is very little flow fluid affecting the motor means that the power produced by the motor will be relatively small. Increasing the angle of the swash plate will allow the pistons to move farther. This means the more fluid acting on the pistons, the more power the motor can generate, and yet operate at a slower speed.

Bidirectional Motors with Pressure Compensation

Adding pressure compensation **(Figure 4-16)** to a hydraulic motor allows the motor to increase or decrease in speed as the pressure being applied to the motor changes. Generally, as the pressure increases, the angle on the swash plate will also increase. As the angle on the swash plate increases, the displacement of the pistons also increases, therefore the speed of the motor decreases but the power capacity of the motor increases.

PARTIAL REVOLUTION OSCILLATORS

An oscillator is a device that will move through an arc of less than 360 degrees, and when it reaches the end of this travel it automatically reverses and travels through an arc in the opposite direction. A partial revolution oscillator can be thought of as a hydraulic motor that only makes a partial rotation in one direction before reversing on itself and making a partial rotation on the opposite direction **(Figure 4-17)**.

Figure 4-17 A partial revolution oscillator.

CONTROL VALVES

Control valves are the nerve center of any fluid power system. The cylinders and motorists may do the work, but it is the control valves that determine what work will be done, when it will be done, and how it will be done.

There are several terms to be familiar with before looking at valve symbols. A two-way valve is a valve that has only two hose fittings connected to it: one inlet and one outlet. In a three-way valve, three ports exist for hydraulic fluid flow. At any given time, one of these ports may be an inlet and the other two outlets, or vice versa. Most control valves have a position that is considered their normal position. This is the position to which the valve returns when no forces are being applied to make the valve move to a specific position. This normal position may be where all the valves are open or closed; if the valve is normally open and fluid can flow through the valve, it's in the rest position. If the valve is a normally closed valve, it will stop fluid flow while in its rest position.

Two-Way, Two-Position Normally Closed Valves

A two-way, two-position normally closed valve is simply an on/off switch with the rest position being the off position **(Figure 4-18)**. When moved to the non-rest position, fluid can flow freely through the valve from the pump to the actuator. When released from control, the normally closed valve will return to its rest position, which will shut off fluid flow from the pump to the actuator.

Two-Way, Two Position Normally Open Valves

This valve is simply the opposite of the valve discussed previously. When released to its rest position,

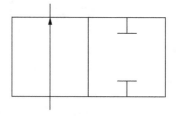

Figure 4-19 A two-way, two-position normally open valve.

fluid can flow freely from the pump to the system's actuator. When the valve is moved to its non-rest position, the flow of fluid is cut off **(Figure 4-19)**.

Three-Way, Two-Direction Control Valves

Fluid power control valves are usually not designed for one specific job and in one specific way. A given valve design can be used for a number of different jobs, depending on how it is placed in the system. For instance, a three-way directional control valve can be used in a system in a number of different ways. Imagine a system with two hydraulic sources supplying two different flow rates to an actuator **(Figure 4-20)**. In most cases, we would not want both flow rates available to the actuator simultaneously. In one valve position, flow from source A would be shut off at the valve, while flow from source B would be allowed to continue on to the actuator. If the valve were moved to the other position, flow from source B would now be cut off and flow from source A would be sent to the actuator.

Three-Way Selectors

The three-way selector **(Figure 4-21)** is used when there is a single fluid power source but two actuators

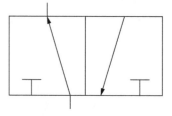

Figure 4-20 A three-way, two-direction control valve.

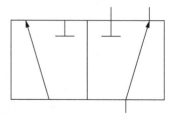

Figure 4-21 A three-way selector.

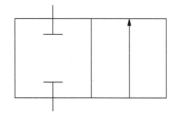

Figure 4-18 A two-way, two-position normally closed valve.

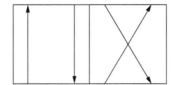

Figure 4-22 A four-way, two-position single actuator.

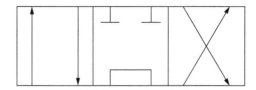

Figure 4-24 A four-way, three-position tandem center valve.

from which to choose. In one position, the fluid will flow to actuator A, and in the other position the fluid will flow to actuator B. In both cases, the actuator not receiving the flow of fluid will be closed off so the fluid in the actuator will not return to the reservoir.

Four-Way, Two-Position Single Actuators

Many hydraulic applications require a cylinder or motor to move in one direction and then reverse and move in the other action. The four-way, two-position single actuators valve has two inlet ports and two outlet ports **(Figure 4-22)**. Hydraulic fluid can pass through the valve on its way to the actuator, pass through the actuator, and then pass back through the directional control valve. When the valve slides to the other position, the hydraulic fluid then passes through the actuator in the opposite direction and also returns in the opposite direction. This allows for reversing the direction of operation of the actuator.

Four-Way, Three-Position Closed Center Control Valves

Although it is impossible to truly say anything is "the most common" type of device in a hydraulic system, the four-way, three-position closed center directional control valve comes close **(Figure 4-23)**. In the center position, flow cannot occur from the pump, flow cannot occur back to the reservoir, and flow cannot occur in either direction for the actuator. In the straight through flow position, flow can proceed from the pump to the actuator, through the actuator, return back through the control valve, and then reach the reservoir. When the valve is moved to the cross flow position, the supply path and the return path cross inside of the valve and the direction of the actuator reverses.

Four-Way, Three-Position Tandem Center Valves

In many applications, it will be necessary to be able to provide flow and reverse flow to an actuator while at the same time being able to park that actuator. While the closed center directional control valve is capable of doing these three tasks, the closed center also dead heads the pump. If the system is designed with adequate pressure relief, this is not a problem. However, it does mean that the pump house must work up to the **relief valve** pressure, thereby making the pump work harder than needed. By combining the open center on the pump side of the directional control valve, the tandem valve allows the actuator to park while letting fluid flowing from the pump simply return to the reservoir **(Figure 4-24)**. This decreases the load on the pump, decreases the load on the pressure relief valve, and also decreases the amount of heat in the system.

Four-Way, Three-Position Float Center Valves

Many pieces of equipment require a "float" mode of operation. Float allows fluid to shuffle back and forth between the extend side and retract side of a cylinder or between the forward and reverse sides of a hydraulic motor. At the same time, fluid can return to the reservoir since the return line to the reservoir is also open.

The float feature is important on applications like snowplows. When removing heavy snow and ice it is important for the plow blade to be able to cut into the snowpack or into the ice. When the snow is light or thin on the surface of the road, the blade needs to skim across the pavement; digging in could damage the road surface. Under these conditions, putting the snow plow into float mode allows the blade to skim lightly across the surface of the road, removing the light snowpack. The float position of the four-way, three-position float center valve allows this to happen **(Figure 4-25)**.

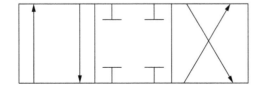

Figure 4-23 A four-way, three-position closed center valve.

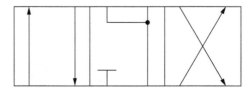

Figure 4-25 A four-way, three-position float center valve.

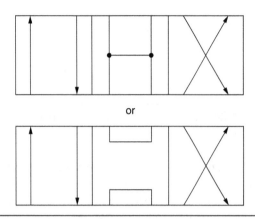

or

Figure 4-26 A four-way, three-position open center valve.

Four-Way, Three-Position Open Center Valves

The four-way, three-position open center valve allows the fluid to flow freely both in a forward direction and in a reverse direction **(Figure 4-26)**. With the valve brought to the center position, both the supply/return side of the valve and the actuator side of the valve will allow fluid to flow freely. This means the fluid from the pump can flow freely and easily through the valve and back to the reservoir. This puts minimal load on the pump and generates the least amount of heat. Fluid on the actuator side of the valve can move freely from the extend side to the retract side of the cylinder, or from the forward side to the reverse side of a hydraulic motor.

VALVE CONTROL SYMBOLS

Each of the control valve symbols have modifiers drawn on it to specify what controls it, how it is controlled, and how it returns to its neutral position. These symbol modifiers can tell the technician a great deal about how hydraulic systems and their controls are designed to behave.

General (Unspecified) Controls

Two parallel lines with a bar across their top or end indicate the valve does have a control system but is not specific about the method of control **(Figure 4-27)**. This symbol can indicate that the valve is manually operated, electrically operated, air operated, or operated by some other undesignated method. In most cases, however, the symbol indicates that the directional control valve is manually operated.

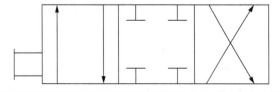

Figure 4-27 A general (unspecified) control.

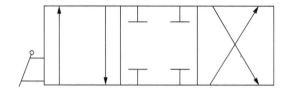

Figure 4-28 A manual lever control.

Manual Lever Controls

Parallel lines with a diagonal line extending across the bars and a circle on one end denote a directional control valve that is operated by a manual lever **(Figure 4-28)**. This symbol is used when the valve is controlled directly by the hands of the operator.

Foot-Operated Controls

Parallel lines with the diagonal line extending across the bars and a 90 degree hook on one end indicate a foot-operated control valve **(Figure 4-29)**. The symbol is used in situations where the directional control valve or other control valve is controlled with either the operator's foot or knee.

Cam-Operated Controls

Two parallel lines with a circle across the end of the lines indicate a cam-operated control valve **(Figure 4-30)**. The circle represents the roller that would be riding on the cam to operate the valves.

Pilot-Operated Controls

A pilot-operated control valve **(Figure 4-31)** is one that changes position as a result of a small amount of

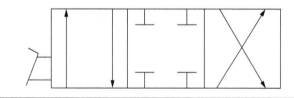

Figure 4-29 A foot-operated control.

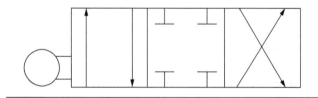

Figure 4-30 A cam-operated control.

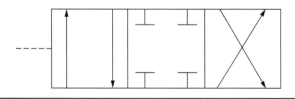

Figure 4-31 A pilot-operated control.

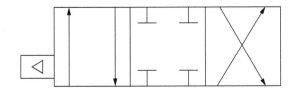

Figure 4-32 A button bleeder control.

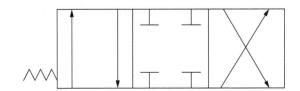

Figure 4-34 A spring-return control valve.

hydraulic fluid under pressure being directed to a chamber within the valve, which causes the valve to shift position. Pilot operation of any valve is indicated by the dashed line drawn entering the valve at one end. The directional control valve can be pilot-operated in one direction or in both directions.

Button Bleeder Controls

Some hydraulic systems have a need to purge small amounts of hydraulic fluid by hand. This is done using a valve that can be pushed open with either a finger or a thumb **(Figure 4-32)**. Such a thumb-operated valve could also be used to bleed off a small amount of hydraulic fluid, the pressure of which is being used to hold a valve in a specific position. When the pressure is bled off, it allows the valve to move to the center or reverse flow position. Therefore, the bleeding of a small amount of fluid using a thumb-operated valve can stop flow through the valve and reverse the direction of flow through the valve. This method of control is called button bleeder control.

Solenoid-Operated Control Valves

Most of today's fluid power systems are operated in conjunction with electrical or electronic controls. One of the most common devices used to link electrical control to the fluid power system is the solenoid-operated valve **(Figure 4-33)**. The symbol that indicates a valve is solenoid operated is a rectangle with a diagonal line through it. The symbol designates that when the solenoid is activated, the valve is pulled in the direction of the solenoid.

Spring-Return Control Valves

The control symbols discussed so far represent devices that are used primarily to move the valve from its center or rest position. Although any of the aforementioned controls could be used to center the

valve, the most common centering device is a spring **(Figure 4-34)**.

Pressure-Compensated Control Valves

The control valve is pressure-compensated **(Figure 4-35)** and will respond to changes in the system pressure, moving the valve to a predetermined position based on system design. One use of this type of valve would be in a winch to prevent overloading the device. As the load on a hydraulic motor increases, the pressure in the system upstream of the motor also increases. If the motor operating the winch is loaded to the point where the pressure in the system becomes too high, the pressure-compensated valve would return to the center position and terminate the lift. Depending on what additional valves or components are in the system, the winch motor may simply park until the operator gives the system an additional command, or the system may be designed to automatically enter a self-protection mode.

Pilot and Solenoid-Control Valves

As previously mentioned, pilot control uses a small flow of hydraulic fluid to control the movement operation of the valve. Stacking an electrical solenoid along with the pilot control valve allows an electrically controlled system to determine when pilot pressure can be applied to the valve **(Figure 4-36)**. This can be designed to operate in either an "AND" or in an "OR" capacity. When set up in the "AND"

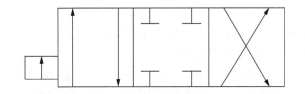

Figure 4-35 A pressure-compensated control valve.

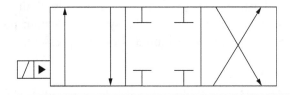

Figure 4-36 A pilot and solenoid-control valve.

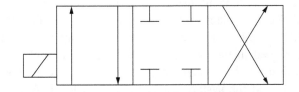

Figure 4-33 A solenoid-operated control valve.

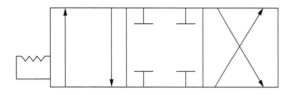

Figure 4-37 A three-position detent.

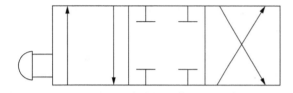

Figure 4-38 A palm button control.

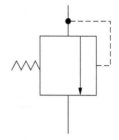

Figure 4-39 A relief valve.

configuration, both the pilot pressure and the electrical current must be flowing through the solenoid in order for the valve to be moved. When set up in the "OR" configuration, either the solenoid must be electrically activated *or* there must be a pressure applied to the control of the valve for the valve to move.

Three-Position Detents

The symbol representing the three-position detent **(Figure 4-37)** signifies that the spool of the valve does not move back and forth through the valve freely; rather, it locks tightly in place in each position. This is usually accomplished with a spring-loaded bowl that rides over ridges on a shaft that is connected to the spool valve.

Palm Button Controls

Palm button controls **(Figure 4-38)** are usually used where urgent action may be required on the part of the operator. One very common place where these controls are used is on an emergency shutoff valve. With this type of control, should an emergency present itself, all the operator has to do is simply reach out and slap the button.

PRESSURE REGULATING, PRESSURE LIMITING, AND SAFETY VALVES

Fluid power systems inherently have a great deal of potential energy and therefore a high potential for danger. Designed into all fluid power systems are safety valves, pressure limiting valves, and pressure-related valves. These valves automatically limit the amount of pressure within the system or within portions of the system to limit and control pressure.

Relief Valves

The pressure relief valve is one of the most common automatically actuating valves in fluid power

systems **(Figure 4-39)**. In almost all fluid power systems, the pressure relief valve will be located parallel to the output of the hydraulic pump. This ensures that should the flow of the highlighted hydraulic fluid be deadheaded anywhere downstream of the main pump output line, the pressure in the outline and the pump will not exceed a predetermined amount. If this valve is not there, and should the flow of hydraulic fluid reach a dead end, or in other words become deadheaded, the hydraulic pressure in the pump output line would reach the internal bypass pressure of the hydraulic pump. This at the very least would cause the internal temperature of the pump to rise to a damaging level. In a worst-case scenario, the pump output hose could rupture, spraying hot hydraulic fluid for a significant distance.

The pressure relief valve consists of a spring-loaded spool held in a normally closed position. The pilot line, usually in the form of a passageway inside the valve, supplies hydraulic pressure to the opposite side of the spool. When the pressure in the pilot line becomes sufficient to offset the tension on the spring, the position of the spool valve will shift, allowing fluid to flow through the valve. In most cases, the spring tension is extremely high. This means that the pilot line pressure would need to exceed the normal operating pressure of the hydraulic system before the spool can shift position and allow fluid to flow, bypassing the rest of the hydraulic system. The end result is that a relief valve, even in the worst case, would prevent the system from overpressurizing to a damaging level.

Many people new to fluid power systems make the mistake of thinking that every type of component, such as a relief valve, is designed to do a specific job, and *only* that job. This may be true to some extent, however a component designed to be a pressure relief valve can be placed in the system to do a significantly different job. For instance, a pressure relief valve can be placed in the system to prevent flow through one portion of the system until the pressure in another portion of the system reaches a certain level. At that point, the relief valve will open and allow fluid to flow

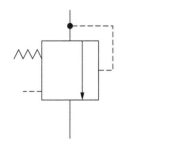

Figure 4-40 A relief valve with a vent.

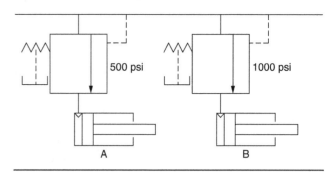

Figure 4-42 A sequence valve diagram.

into the previously deprived portion of the system. In this case, the relief valve would no longer be a pressure relief valve but rather would be sequencing fluid flow through different portions of the system based on pressures. This would therefore be called a sequencing valve. Always keep in mind that what the valve is doing in the system determines what the valve will be called; it is not the design that determines what a valve will be called.

Relief Valves with Vents

Adding a vent to the pressure relief valve allows the hydraulic fluid that is contributing to the over-pressurization to escape the system **(Figure 4-40)**. Typically, this device would only be used when the pressure relief valve is, of necessity, remotely located from the reservoir. Hydraulic fluid escaping through the vent would eventually collect on the ground creating a safety hazard or perhaps an environmental hazard. For this reason, vents are seldom used on modern fluid power systems.

Sequence Valves

Sequence valves (Figure 4-41) can be employed in many ways. A typical example would be the hatch and ram system employed in some refuse trucks. Many of these trucks employ a large plate on a hydraulic ram to first compact the trash and later push the trash out of the hopper. When the trash is being compacted, the hatch remains closed and the trash is compacted against the hatch. A pressure relief valve is employed during compacting to prevent the ram from pushing

the trash against the door with enough force to damage the door. A different hydraulic circuit is employed to push the trash out of the hopper. The hydraulic fluid going to the ram is routed with a different pressure relief valve to a sequence valve. The sequence valve stops the fluid from flowing to the ram until a certain pressure is reached. This one is set to a higher pressure so the ram can employ more force to remove the contents of the hopper. This higher force is great enough to damage the hatch; therefore the hatch must be already open when the ram begins to push. A cylinder with a relatively large surface area on the piston is employed to open the hatch. The large surface area allows the cylinder opening the hatch to open it at a relatively low pressure, a pressure well below what is required to open the sequence valve that is preventing the fluid from going to the ram. Therefore, as the pressure rises, the hatch opens, reaching its full extension, and since no more movement is taking place, the pressure begins to rise, eventually reaching the pressure that allows the sequence valve to open, which in turn allows fluid to flow to the ram and lets the ram push the trash out of the refuse truck hopper **(Figure 4-42)**.

Pressure Reducing Valves

There are times when fluid in one portion of a system is tapped to operate a parallel device. Sometimes this parallel device has lower pressure limits or pressure requirements. At such time, a **pressure reducing valve (Figure 4-43)** can be used. The pressure reducing valve is similar to a pressure relief valve

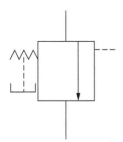

Figure 4-41 A sequence valve.

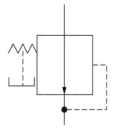

Figure 4-43 A pressure reducing valve.

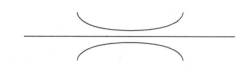

Figure 4-44 A fixed orifice.

except for one fundamental difference. The pressure reducing valve is a normally open valve, when the pressure in the circuit reaches a certain point, the valve closes, thereby limiting pressure downstream of the valve. There is a drain line that allows excess flow to return to the reservoir.

Fixed Orifices

The most common way to limit the volume of flow into any portion of a fluid power system is to pass the fluid through a restriction such as a fixed orifice **(Figure 4-44)**. As discussed in previous chapters, the volume or flow rate of the hydraulic fluid is dependent on the pressure end of the size of the component through which the fluid is passing. The smaller the passage, or in this case the orifice, the smaller the volume of fluid that can flow through the orifice during any given time period. An orifice can be used to slow down the movement of the piston, the cylinder, or the rotating speed of a hydraulic motor. More than one orifice of different diameters can be used in parallel, so the speed at which the piston or motor operates can be varied based on which orifice is selected for the fluid to pass through on its way to the actuator.

Needle Valves or Variable Orifices

Some fluid power systems have actuators that may need to work at different speeds, depending on the operator's judgment. In these situations, a needle valve or variable orifice is used **(Figure 4-45)**. This allows the operator to change the position of the adjustment on the needle valve and alter the flow rate of the hydraulic fluid through the system.

Pressure-Compensated Flow Control Valves

In some fluid power applications, it may be necessary to limit pressure in a portion of the system by decreasing flow to that portion of the system. This can be done using a pressure-compensated flow control valve **(Figure 4-46)**. As the pressure downstream of the flow control valve increases, the flow into that

Figure 4-46 A pressure-compensated flow control valve.

Figure 4-47 An accumulator.

portion of the system is decreased proportionally to the increased pressure. The decreased flow rate slows the rise in pressure.

ACCUMULATORS

Accumulators (Figure 4-47) have two primary functions in a fluid power system. They are used to store a volume of fluid under pressure, and to smooth out pressure pulsations to create a more consistent fluid flow and pressure.

An example where an accumulator might be employed to store fluid is where a hydraulic cylinder is used to open a door at high speed. While the door is closed, the accumulator is charged by the normal flow rate of the system. When the door needs to open, a valve will allow the accumulator to discharge at a high flow rate into the cylinder that operates the door. Depending on design this could allow the door to be opened at many times the normal flow rate of the system at that point.

Hydraulic cylinders, and especially hydraulic motors, can benefit from a smooth even flow rate and a consistent pressure during that flow. Reducing pressure pulses reduces vibration in the system, which results in smoother operation and also reduces noise from the system.

CHECK VALVES

Check valves **(Figure 4-48)** are used to prevent fluid from flowing the wrong way in a fluid power

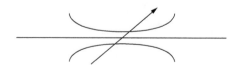

Figure 4-45 A needle valve or variable orifice.

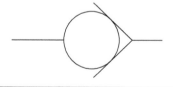

Figure 4-48 A check valve.

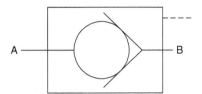

Figure 4-49 A pilot to open check valve.

system. Consisting of a spring load valve; a sufficient pressure differential across the valve can displace the valve from its seat and allow hydraulic fluid to flow through the check valve. When the fluid attempts to move in the opposite direction, the valve is driven hard against its seat and the fluid cannot flow.

Pilot to Open Check Valves

A pilot to open check valve **(Figure 4-49)** allows fluid to flow from "B" to "A" like a standard check valve. Below the pilot pressure, it prevents fluid from flowing from "A" to "B." In most cases, the pilot pressure of these valves is not a specific pressure, but rather a percentage or ratio of the pressure at "B." For instance, if the specification sheet of the pilot to open check valve in question were 3.8:1, then with 1,000 psi at "B," the pilot pressure required to open the valve would be about 265 pounds. If the pressure at "B" were 2,000 psi, then the pilot pressure would be 530 psi.

Most technicians will never need to worry about pilot ratios. In most cases, the schematic or service data for the system being diagnosed will have pilot pressure specifications. The only time it may become an issue is when trying to get replacement parts through a non-standard supply chain.

Pilot to Closed Check Valves

As the name implies, a pilot to closed check valve **(Figure 4-50)** will allow fluid to flow from "B" to "A" but will not allow fluid to flow from "A" to "B." When fluid is flowing from "B" to "A," if the pressure on the pilot port rises above the ratio pressure, then the valve will close to flow in either direction.

Flow Control Valves with Bypass

There are applications in fluid power systems where the cylinder or motor will need to operate at a controlled and limited speed in one direction, yet in the

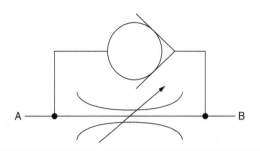

Figure 4-51 A flow control valve with bypass.

other direction control of the speed is not nearly so critical. In a flow control valve with bypass **(Figure 4-51)**, the check valve of the bypass remains closed as the fluid passes through to operate a cylinder or motor in one direction—in this example, "A" to "B." The only path to the actuator is through the flow control valve. The flow control valve can then be adjusted to achieve a desired speed for the actuator. When the directional control of fluid flow reverses "B" to "A," the fluid flows in the opposite direction through the flow control valve, but it also opens the bypass check valve to open, allowing a less regulated and increased flow of fluid. The cylinder or motor then reverses direction and moves at a higher rate of speed.

Pressure-Compensated Flow Control Valves with Bypass

In a pressure-compensated flow control valve with bypass **(Figure 4-52)** type of valve, the fluid flows from "A" to "B" through the flow control valve. However, when flowing from "A" to "B," it does not go through the check valve. When the pressure on the "A" side of the valve changes, the flow rate through the flow control valve also changes. This would allow a cylinder or motor to operate in one direction at a relatively slow and controlled speed. The speed in fact would change as the pressure on the upstream side of "A" changes. When the flow is reversed and allowed to flow from "B" to "A," the check valve will open and allow the cylinder or motor being serviced by this valve to move in the opposite direction at a greater, but less controlled, speed.

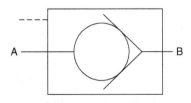

Figure 4-50 A pilot to closed check valve.

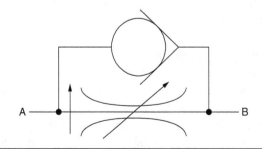

Figure 4-52 A pressure-compensated flow control valve with bypass.

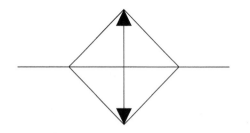

Figure 4-53 A heat exchanger (cooler).

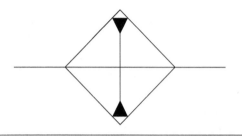

Figure 4-54 A heat exchanger (heater).

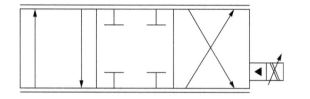

Figure 4-55 A four-way servo valve.

Figure 4-56 A pressure gauge.

To actuator

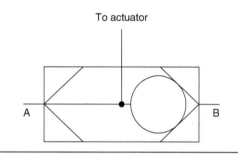

Figure 4-57 A shuttle valve.

Heat Exchangers (Coolers)

Hydraulic fluid becoming pressurized, flowing through small openings, and components, gains heat. This heat can be dissipated in the reservoir if given a sufficient amount of time to rest in the reservoir before beginning a new journey through the system. In many high-cycle-rate systems, an oil cooler will be used to limit the maximum temperature of the oil **(Figure 4-53)**. These coolers can be hydraulic fluid to air or hydraulic fluid to a coolant liquid.

Heat Exchangers (Heaters)

Some fluid power systems have to operate in cold climates or environments. Fluid power systems designed for use in a cold environment will often have a heater **(Figure 4-54)**. This can be a heating element attached to the reservoir or set inside the reservoir. This type of heater is very important in making sure the fluid is ready to flow when the system is inactive. They are largely ineffective in an operating system. To ensure the oil is a stable operating temperature, an inline heater is used. These heaters can be electric, but often the heat comes from another heated liquid, such as engine coolant, or from heated air.

Four-Way Servo Valves

The M1A1 Abrams Main Battle tank features a rather amazing technology in its hydraulic system. When the electronic targeting system is locked onto a target, a hydraulic system is able to keep the gun at the proper angle to the ground for targeting. It is able to do this regardless of the angle of the chassis of the tank. This is done with a servo valve **(Figure 4-55)**.

An electronic control system operates a pair of solenoids that move a directional control valve back and forth. Servo valves allow for rapid and precise control of hydraulically operated systems.

Pressure Gauges

Pressure gauges **(Figure 4-56)** are often placed at critical points in a fluid power system. These may be direct read mechanical gauges or they may be electronic sensors that pass signals about pressures to an electronic control unit.

Shuttle Valves

Shuttle valves **(Figure 4-57)** are used to allow hydraulic fluid to come into an actuator from one direction or the other, but never from both directions at the same time. In the example, fluid flow into the "A" port must be at a higher pressure than the fluid that could flow from "B" port. The higher pressure entering at "A" causes the valve to shuttle and block off the "B" port. Flow from "A" can then flow to the actuator. If the pressure at "B" should rise above the pressure at "A," then the valve will shuttle to block off "A" and flow to the actuator can only come from "B."

Figure 4-58 A manual shut-off valve.

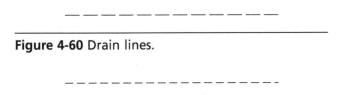

Figure 4-59 A component group outline.

Manual Shut-off Valves

Manual shut-off valves **(Figure 4-58)** are used to manually isolate portions of a fluid power system. These are often placed to facilitate service of components in the system or to isolate a section of a system that has developed a leak. In some cases, where systems are very simple or can operate in a steady state for an extended period of time, manual valves can be used as the operational control valve of the fluid power system.

Component Group Outlines

A combination long-dash, short-dash line surrounding a group of components indicates that the contained components are located within a single assembly **(Figure 4-59)**.

Drain Lines

A line of long dashes is used to indicate a drain line **(Figure 4-60)**. Drain lines are used to carry fluid from non-actuator components back to the reservoir.

Pilot Lines

A series of short dashes forming a line is used to represent lines that carry fluid to a control valve to make the control valve alter its current state of operation. These are called pilot lines and carry fluid to pilot-operated valves **(Figure 4-61)**.

Lines Crossing

When two lines cross on a diagram and there is a dot at the point where they cross, it indicates that the lines, or hoses, are connected **(Figure 4-62)**.

Lines Connecting

If there is no dot where lines cross on a diagram, it indicates there is no connection between the lines or

***

Figure 4-60 Drain lines.

***

Figure 4-61 Pilot lines.

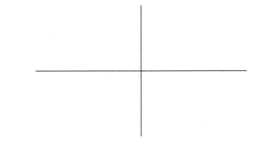

Figure 4-62 Lines crossing.

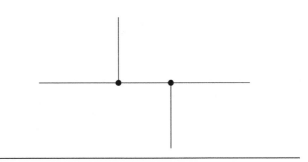

Figure 4-63 Lines connecting.

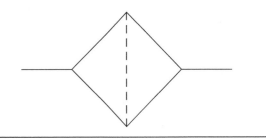

Figure 4-64 A filter.

hoses **(Figure 4-63)**. In these cases, the lines simply cross one another as a convenience to the technical artist making the drawing.

Filters

A diamond placed across a line with a dashed line positioned perpendicular to the line indicates a filter **(Figure 4-64)**.

Building a Simple System

The simplest systems consist of an actuator, such as a cylinder, a reservoir, a pump, and a control valve **(Figure 4-65)**. In this example, the pump draws fluid from the reservoir and the flow is toward the control valve. This control valve is a closed center. When the fluid arrives at the control valve, the closed center provides no path for flow. This will cause pressure to increase, perhaps to the point of damaging the pump or causing a catastrophic leak quite rapidly. We will address this problem later.

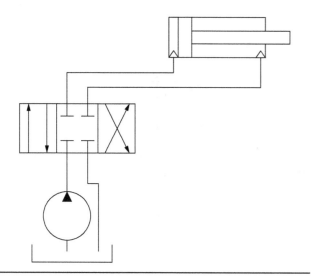

Figure 4-65 A directional control valve in the center or neutral position. The piston in the cylinder is not moving.

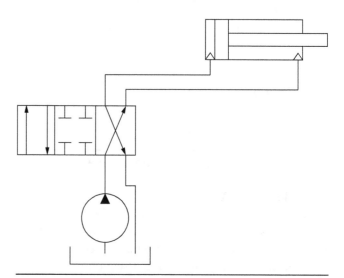

Figure 4-67 A directional control valve in the cross flow position. The piston in the cylinder is retracting.

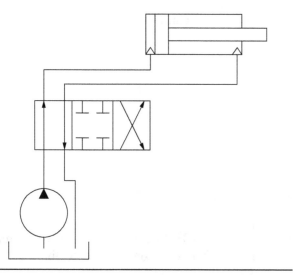

Figure 4-66 A directional control valve in the straight flow position. The piston in the cylinder is extending.

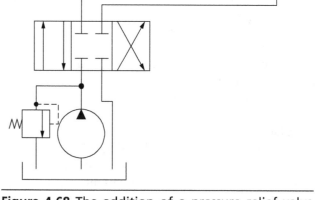

Figure 4-68 The addition of a pressure relief valve protects the pump in the event of a restricted line or when the control valve is in the closed center position.

When the control valve is moved to the straight flow position, fluid can flow from the pump to the extend side of the piston in the cylinder **(Figure 4-66)**. The fluid in the retract side of the piston can then move through the control valve back to the reservoir.

Sliding the directional control valve to the reverse flow position causes the fluid to flow from the pump to the retract side of the piston, causing the piston to move in the retract direction **(Figure 4-67)**. Fluid from the extend side is then forced out of the extend side, through the control valve, and back to the reservoir.

Earlier, it was pointed out that when the valve was in the closed center position, this would cause

pressures to rise to destructive levels. When the directional control valve is in the center position, the pump will push fluid against the closed valve. Pressure will build rapidly. Since there is no place for the pressure to be relieved, a catastrophic failure is likely. Adding a pressure relief valve will accommodate the closed center design of this system **(Figure 4-68)**. When pressure gets to the relief point, the pressure relief valve opens and fluid can flow back to the reservoir, thereby limiting the pressure.

When the flow control valve with a bypass is added to this system, the piston will move at a controlled speed in the extend direction **(Figure 4-69)**. When the directional control valve is moved into the retract

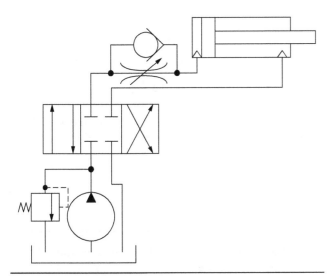

Figure 4-69 When a flow control valve with a bypass is added, the speed at which the piston extends is limited and controlled. When the directional control valve reverses flow, the flow is increased.

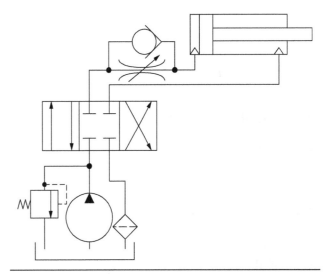

Figure 4-70 A filter is a common component mounted on the return line into the reservoir.

position, flow will reverse direction. Through the flow control valve, the flow in the retract direction will be the same as it was in the extend direction. However, in the retract direction, the check valve that lies parallel to the flow control valve will open. The flow rate of the fluid leaving the extend side of the cylinder can now increase. If the system is designed to allow it, the flow rate into the retract side can now increase because the flow out of the extend side has increased. Thus, the piston will retract at a higher speed than when it is extended.

Cleanliness is essential in a hydraulic system. Adding a filter on the return side of the system where the return fluid enters the reservoir is common practice **(Figure 4-70)**. These filters are usually made of a cellulose material or a synthetic material. They are placed on the return line because it ensures that all the contamination picked up by the fluid is filtered out before the contamination reaches the reservoir. It is also easier for the fluid to be pushed through the filter by the pump than it is for the pump to pull the fluid through the filter from the reservoir.

By adding two solenoid-operated valves, an orifice, and an accumulator, the system is modified to allow for the maximum speed in retracting the cylinder **(Figure 4-71)**. During normal operation, during the extend or retract position, or even while idle, the accumulator is charging up to system pressure. When the operator decides that a rapid retraction of the piston is required, the two solenoid-operated valves open. Fluid flows rapidly out of the accumulator; the new orifice prevents the fluid from flowing back through the return line and exceeding the flow rating of the

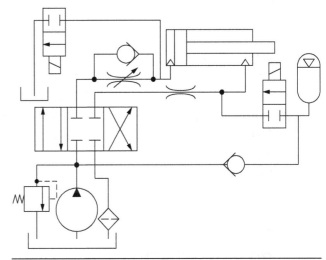

Figure 4-71 The addition of an accumulator, a check valve, and a pair of solenoid-operated two-way, two-position normally closed valves allows an electrical control to use a flow of fluid from an accumulator to greatly, and suddenly, increase the speed of the piston when retracting.

filter. The bulk of the fluid flows rapidly into the retract side of the cylinder. As the piston is forced to retract, it forces fluid from the extend side back through the flow control valve and check valve. However, because the solenoid-operated valve is also open, large amounts of additional fluid can flow back to the reservoir. The result is a very rapid retraction of the piston within the cylinder.

POWER SOURCES

Many hydraulic system drawings do not illustrate the power source for driving the pump. This is

Figure 4-72 An internal combustion engine.

Figure 4-73 An electric motor.

because the designer of the fluid power system does not always know what the power source will be. A square with an "M" in it designates the power source as an internal combustion engine **(Figure 4-72)**. A circle with an "M" in it represents an electric motor **(Figure 4-73)**.

Summary

Any technician involved in the maintenance or repair of fluid power systems on mobile equipment needs to become very familiar with the symbols used in the manufactures diagrams. The symbols in these diagrams describe how the system functions and what is expected of the system when it functions properly.

Each one of these symbols tells a story to the trained technician, which would literally take hundreds or thousands of words to describe. The more experienced a technician becomes at understanding and interpreting the stories told by the symbols, the more successful he will be in troubleshooting.

Review Questions

1. The North American organization that governs symbols used in fluid power systems is the:

   A. American National Standards Institute.

   B. Society of Automotive Engineers.

   C. National Institute for Automotive Service Excellence.

   D. National Fluid Power Society.

2. A symbol consisting of a square with a capital "M" inside of it designates that:

   A. Power for the system is supplied by an electric motor.

   B. The system has been modified from its original production design.

   C. Power for the system is supplied by an internal combustion engine.

   D. A meter is installed at that point.

3. Which symbol represents a variable displacement motor with pressure compensation?

4. Small numbers next to a symbol may:

   A. Represent an opening pressure.

   B. Represent a maximum pressure.

   C. Represent a minimum pressure.

   D. Have no meaning.

5. Federal law requires that all hydraulic system manufacturers use a specific set of symbols.

A. True   B. False

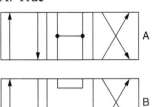

6. Which of the preceding symbols represents a four-way, three-position open center directional control valve?

A. A   C. Both A and B

B. B   D. Neither A nor B

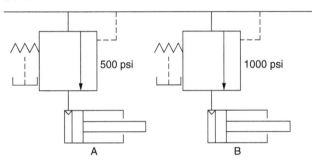

7. As pressure builds in the preceding circuit, which piston will begin to extend first?

A. A   C. They will both extend simultaneously.

B. B   D. The circuit shown will only retract.

8. 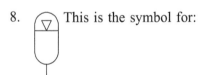 This is the symbol for:

A. A double check valve.   C. An accumulator.

B. A hydraulic solenoid.   D. A tapered valve.

9. A shuttle valve:

A. Allows fluid to come from only one source to a given section of a system.

B. Allows fluid to only come from the source with the highest pressure.

C. Both A and B

D. Neither A nor B

10. An accumulator can be used to:

A. Smooth out variations in flow.

B. Smooth out variations in pressure.

C. Supply a large burst of pressurized hydraulic fluid for the rapid operation of a motor or cylinder.

D. All of the above

E. None of the above

CHAPTER

5 Oils and Other Hydraulic Fluids

Learning Objectives

Upon completion and review of this chapter, the student should be able to:

- Describe the different types of hydraulic fluids and other purposes.
- List the many ways that hydraulic fluids can deteriorate.
- List causes for the deterioration of hydraulic fluid.
- List the additives used in hydraulic fluids, and the purpose of each additive.

Cautions for This Chapter

- Although most hydraulic fluids are harmless when they come in contact with the skin, prolonged exposure can cause skin irritation.
- Always wear gloves that are resistant to the type of oil being handled.
- Always wear a face shield or at least safety glasses when working with hydraulic fluids.
- Spilled oil can be very slick and thus hazardous to walk on or handle equipment on.

Key Terms

acid	corrosion	lubrication
alkaline	entrainment	rust
base		

PURPOSE

Transmission of Force

The primary purpose of hydraulic fluid is to transmit power from one part of the fluid power system to another part. To do this in an efficient manner, the fluid must be non-compressible. When a pipe is filled with fluid and a force is applied to the fluid at one end of the pipe, the fluid must be able to transfer that force to the other end of the pipe without a significant loss of force or delay. Systems that use compressed gases, such as pneumatic systems, must first compress the force transfer medium (for example, air) in order to transfer that force from one point to another. This causes a delay in response with pneumatic-based machinery (**Figure 5-1**). Since there is no tangible compressibility in hydraulic fluid (or most other liquids), there is no delay. When a valve opens,

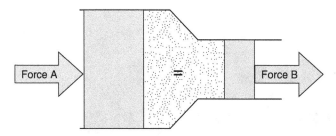

Figure 5-1 Hydraulic fluid is a non-compressible energy transfer medium. This means that when a force is applied to a piston, that same force will be transferred to the second piston. There may be changes in pressure with the transfer medium, which will absorb only a minimal amount of energy to transfer that force.

allowing high-pressure fluid to flow against a lower-pressure fluid, than the increase in pressure on the components having the lower-pressure fluid is immediate. Additionally, the active compressing of the pressure medium, such as air, expands energy. This energy will not be transferred to the actuator. Therefore, an energy loss occurs. Because of the non-compressibility factor, hydraulic systems can operate at very high speeds with very little energy loss.

LUBRICATION

The second purpose of the fluid is **lubrication**. The lubrication properties of the fluid must help protect system components from friction and wear. Since the bulk of the components and subcomponents are metallic **rust**, oxidation and **corrosion** are also a possibility. The lubricating properties of the hydraulic fluid coat the components to reduce friction as they slide past one another, thereby reducing heat. These same properties result in a film of the fluid being retained by the components that reduce rust, oxidation, and corrosion **(Figure 5-2)**. As the hydraulic oil ages and the additives break down, and the oil becomes infiltrated by air and moisture, its ability to form and maintain a film between sliding surfaces degrades. How long this takes depends on operating conditions, temperatures, and the environment.

Oil film

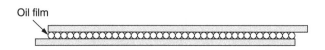

Figure 5-2 Hydraulic fluid also serves as a lubricant. Even finely machined metal surfaces will develop friction sliding against one another. This friction is greatly reduced when there is a film of oil—or in the case of a fluid power system, hydraulic fluid—separating two machined metal surfaces.

SEALING

Although the components in a fluid power system are machined to very close tolerances, minute gaps exist between the metal components. Rubber or rubber-like seals are impractical because of the high pressures involved. The retained film of hydraulic fluid on the metallic components acts to form a seal. The ability of a given formulation of oil to do this is dependent upon the physical characteristics of the hydraulic fluid. This is why it is essential that only oils approved by the fluid power system's manufacturer be used in a given system.

All liquids have a characteristic known as surface tension. If you have ever seen an insect walking on water, or ever noticed the meniscus formed when water or another liquid is retained in a small tube, then you have seen the effects of surface tension **(Figure 5-3)**. The surface molecules of the liquid hold together very firmly. In each case, it takes a certain force to break the hold. Since the insect is relatively light, it is able to walk across the water because its weight does not apply enough force to break the surface tension.

This principle allows oil coating the surfaces of hydraulic systems, especially valves, to form a barrier along the surface of the film. This barrier caused by the surface tension forms a seal. In general, the greater the viscosity of the oil, the greater is the ability to form a seal. As the oil ages, viscosity tends to break down, and therefore the ability of the hydraulic oil to form a seal also degrades.

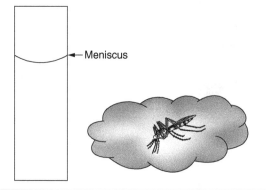

Figure 5-3 The surface of any pool or container of liquid has a characteristic called surface tension. This is the characteristic that forms the depression along the surface of water held in a tube. It is also the characteristic that allows insects to walk along the surface of the water. The weight of the insect is not sufficient to break the bonds that form the surface tension. Therefore, the insect can easily walk along the surface in exactly the same way that humans walk along the surface of the sidewalk. Surface tension allows hydraulic oil to form a seal to prevent not only that oil but other oil from leaking.

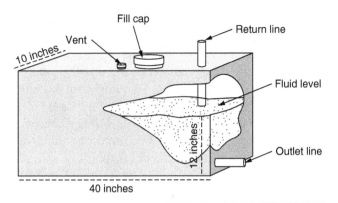

Figure 5-4 Many applications use a liquid-to-air heat exchanger or a liquid-to-liquid heat exchanger to ensure that the heat picked up by the fluid as it passes through the various components of the system have a place to give up that heat. Simpler applications use the reservoir as a chamber when the fluid is allowed to stand, giving up heat to the walls of the reservoir, which then passes the heat onto the surrounding air. By the time the oil is picked up again by the pump, it has had an opportunity to cool and is ready to absorb heat as it passes through the fluid power system components once again.

COOLING

The hydraulic fluid is also essential in removing heat from the metal components of a fluid power system. Different hydraulic fluid oils have different heat absorption and dissipation rates. If the wrong oil is used, or if the correct oil is diluted with an improper fluid, then the ability to absorb, transport, and properly dissipate heat can be reduced.

Hydraulic fluid uses a wide variety of **base** oils or other liquids. Anyone who's ever sat near a lake at evening knows that the land cools before the water of the lake does. Liquids tend to repel heat when they are cool and tend to hold on to heat when they are warm. Different liquids have different characteristics when it comes to absorbing and giving up heat. Since most components in a hydraulic system have little or no extra cooling, the typical system will depend on the circulation of hydraulic fluid to remove heat from the components and give up that heat while the fluid is resting in the reservoir **(Figure 5-4)**.

Physical Characteristics

VISCOSITY

Viscosity is the property of a fluid, such as hydraulic fluid, to resist the force that is causing the fluid to flow. Using a fluid with too high of a viscosity rating can increase system pressures and cause

Figure 5-5 Honey clings to this honey-dipping tool because it is a high-viscosity liquid. Although the tool works wonderfully with honey, one can only imagine how inefficient it would be to attempt to transfer water using this tool.

excessive pressure drops at every restriction to flow in the system. This also decreases flow rate. Since any restriction to flow generates heat, high-viscosity hydraulic fluid can increase the amount of heat generated by the fluid flowing through the hydraulic system **(Figure 5-5)**. As might be expected, if the hydraulic oil chosen has too low of a viscosity, its ability to stick to metal components is lower. As a result, its ability to lubricate and protect from corrosion is also decreased. Poor lubrication can cause excessive wear and even seizure of components.

An example of a liquid with low viscosity is water. Honey is an excellent example of a substance with high viscosity. The water has very little resistance to flow and the honey has a very high resistance to flow.

COMPRESSIBILITY

Liquids are used as the force transfer medium in a fluid power system because under most conditions they are noncompressible, and noncompressible liquids provide instantaneous transfer of force. Gases can be used to transfer force, but they are highly compressible **(Figure 5-6)**. For a fluid power system to perform optimally, it is necessary that a force directed to the fluid in a pipe be transferred from the pump end of that pipe to the actuator end of the pipe as rapidly as possible. If the hydraulic fluid in use allows for too much compressibility, there will be a delay in operation as the pressure rises through the pipe from the pump to the actuator. Hydraulic fluids feature a compressibility of approximately one half of one percent per thousand pounds per square inch applied to them, up to about 4,000 pounds per square inch. At pressures above 4,000 psi, compressibility begins to increase with

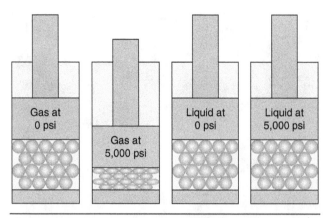

Figure 5-6 When gas is placed under pressure, it is easily compressed. However, when a liquid such as hydraulic fluid is placed under high pressure, the amount of compression is barely noticeable. In many cases, the volumetric loss under pressure is less than one half of one percent.

petroleum-based fluids. Additionally, as the temperature of a petroleum-based fluid increases, the compressibility also increases. High-pressure fluid power systems often require special synthetic hydraulic fluids.

When compressibility becomes an issue, components fail to maintain consistent positioning, there are delays in the operation of cylinders and motors. Compressibility can also result in system cavitation. Cavitation can result in metal fracture, corrosive fatigue, and stress corrosion.

STABILITY

Most hydraulic fluids are petroleum based. As a result, they are prone to deterioration. Even the simple exposure to atmospheric oxygen can degrade the fluid and many of the additives in the fluid. Add heat to the equation and the oxidation process can accelerate. Proper cooling of the fluid, following the recommended service intervals, proper filtration, and other maintenance can reduce the amount of oxidation and its affects on the hydraulic system. Hydrolysis or absorption of water or moisture can dramatically hasten the deterioration of the hydraulic fluid. It is therefore important to maintain the system filters, use only the type of oil recommended by the manufacturer, maintain the fluid at an appropriate temperature, and avoid introducing water or moisture into the fluid.

QUALITY REQUIREMENTS

The quality of hydraulic fluid is predictive of how long the fluid's essential properties will perform according to their manufactured specifications. The essential properties of a quality fluid are oxidation

stability, rust prevention, foam resistance, water separation and anti-wear properties. While some of these properties exist to some level in the typical petroleum-based hydraulic fluid, additives are usually used to enhance these characteristics. One might think therefore that adding additives to hydraulic fluid based on a preconceived notion of how the fluid will perform over time would be a good idea. Unfortunately, additives that enhance the life expectancy of the fluid with regard to one of its properties can often degrade the life expectancy of another of its properties. The additives added by the fluid's manufacturer are a delicate balance of enhancement versus adverse affects. When a technician adds an aftermarket additive, the technician is as likely to reduce some important characteristics as they are to enhance others.

OXIDATION STABILITY

Oxidation is the chemical merger of the hydraulic oil and atmospheric oxygen. Oxidation is also a primary factor in the destabilization of hydraulic fluid, and once the reaction process begins, its effect increases rapidly. Most common hydraulic oils (fluids) are mineral based, or, in other words, derive from petroleum that has been pumped from the ground. This means that one of the primary substances found in hydraulic fluid is a hydrocarbon. The fluid also contains at least trace amounts of sulfur. When oxidation occurs, the hydrocarbon breaks down and combines with the oxygen, yielding carbon dioxide and water. This process alters characteristics of the fluid, including viscosity and flow capability. A worse result of oxidation comes from the trace amounts of sulfur. When sulfur combines with oxygen, hydrogen often joins the party and sulfuric **acid** is formed **(Figure 5-7)**. This can cause rapid deterioration of not just the seals and other soft components of the system, but can damage the hard metal components as well. The affects of oxidation are accelerated by temperature, pressure, contaminants, water, exposure to certain metals, and agitation.

Temperature

Chemical reactions such as oxidation increase as the temperature of the hydraulic fluid increases. High temperatures can cause contaminants to bond with the hydraulic fluid and the chemical compounds that make up the additives of the hydraulic fluid to form new compounds. The effect is similar to the process of cooking. New compounds are formed offering new smells and tastes when the food is heated **(Figure 5-8)**. Although the exact rate at which chemical reactions increase as temperature increases varies with each chemical combination, basically for every 10°C (18°F)

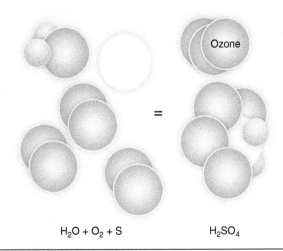

$$H_2O + O_2 + S \qquad H_2SO_4$$

Figure 5-7 All petroleum-based oils have at least a small amount of sulfur in them. Although in the last few years the Environmental Protection Agency has set limits on the amount of sulfur that can be contained in any oil, even in the most modern oils there is at least a trace of sulfur. Sulfur is a very volatile element. It can combine easily with the oxygen from the atmosphere and both oxygen and hydrogen from water to form sulfuric acid (H_2SO_4). Please note that in the example here, ozone was also generated. Ozone itself is known to be a corrosive agent.

Figure 5-8 In much the same way that adding heat to a pan full of food brings the various parts together to produce flavors, textures, and an appeal that goes far beyond the individual parts, when heat is added to hydraulic oil filled with contaminants, unexpected materials can result. Adding heat to hydraulic fluid in the presence of contaminants can produce polymers, abrasives, and compounds that can feed lubrication or the sealing capacity of the oil.

the potential for chemical reaction doubles. In poorly designed or maintained systems, there can be local hot spots. These are the areas where the oxidation process can start and then spread.

Pressure

As the pressure in a system increases, the friction caused by the movement of the fluid increases. This friction generates the heat that can cause oxidation and other chemical reactions. Overworking a fluid power system beyond the tolerance in which it has been designed to operate can cause it to operate at pressures that the system was not built to handle. The result is excessive heat being generated and accelerated oxidation occurring.

Contaminants

Contaminants can enter a system in many ways. One of the most common methods is through the air. Almost all mobile fluid power systems are used in environments that are hostile. Industrial environments, construction sites, and manufacturing facilities that use solvents, paint, and other chemicals all contain volatile airborne components that are hostile to hydraulic fluid and accelerate its deterioration. Some of the contaminant compounds also accelerate sludge formation. A 1 percent concentration of sludge in the hydraulic fluid can double the speed of oxidation.

Metal and Water

Some metals, such as copper, are known to act as catalyst when submerged in hydraulic fluid. This is especially true in the presence of water. Some tests indicate that the amount of acid building up in the hydraulic fluid can be increased and accelerated by as much as twenty times when the fluid is contaminated by both copper and water.

Two specific additives are added to the hydraulic fluid to reduce the oxidation. The first is a chain breaker inhibitor. When oxidation begins, the natural hydrocarbons that are elemental to the fluid begin to form long strings that eventually become the sludge. The chain breaker additives interrupt this process. Another additive is used to reduce or negate the affects of copper and other metals. These additives are known as metal deactivators.

Rust and Corrosion Prevention

Rust is formed by the oxidation of ferrous metals such as iron and steel **(Figure 5-9)**. Air that has been trapped in the hydraulic fluid through agitation contains moisture. That moisture can be condensed out of the air whenever the air-laden fluid is under pressure. This water can be trapped deep within the system and contribute to rust and corrosion throughout the system. Rust inhibitors are added to the fluid to coat internal components of the system and reduce rust.

Figure 5-9 Rust and corrosion form when bare metal is exposed to the oxygen in the atmosphere. The presence of moisture can accelerate the process. Good hydraulic oil will have additives that will slow or inhibit the formation of rust and corrosion.

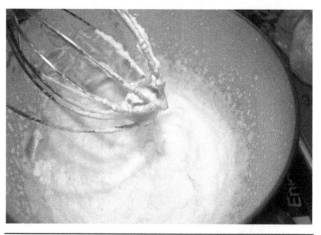

Figure 5-10 Although this looks like a thick, rich meringue, it is actually a liquid into which air has been heartily infused. The egg whites, which are a **base** for this dessert treat, have been entrained with air through a brisk whipping action. Anyone who has ever whipped egg whites and let them stand for a while knows that the air will stay in the egg white until it is allowed to stand for several minutes to several hours. At that point, the air will slowly make its way out and the egg white will turn back into a white liquid. The same process can occur with hydraulic oil. Violent whipping action can blend air into the oil in much the same way as air has been whipped into these egg whites. Just like the egg whites, as long as the air is in the oil, the oil has difficulty moving and flowing.

Agitation

When the hydraulic fluid is agitated, it can capture air. Often this air is captured along the surface of the hydraulic oil. In the worst case, the oil is captured in such quantity that it is blended into foam. This is similar to what happens when air is whipped into egg whites to create meringue **(Figure 5-10)**. This foam is not only compressible but is almost impossible for the pump to pick up from the reservoir. Pump damage from poor lubrication can result. Most hydraulic oils are equipped with an additive to reduce foaming.

Air Entrainment

Air **entrainment** is the capture of very small bubbles throughout the oil. At low pressures typical hydraulic oil without anti-air entrainment additives can absorb up to 10 percent air. As the pressure increases, the ability to absorb air also increases. These bubbles of air expose oil all along their surface area to oxygen. Oil oxidation therefore takes place at a very rapid rate.

The amount of foaming in the fluid is affected by several factors. The viscosity of the fluid and the source of the crude from which the oil was refined can affect its vulnerability to entrained air. Entrained air is particularly destructive to the oil. High pressures in the hydraulic fluid will cause the entrained air to be heated, and then expand, thereby significantly increasing the pressure. The increased pressure brings higher temperatures, which in turn heats the air yet more. The higher temperatures caused by this process can scorch the surrounding oil and cause oxidation.

Like many products designed to relieve a crisis, the overuse of anti-foaming agents can create a secondary problem. The same function of the additive that keeps

air bubbles from forming in the hydraulic oil can also make it difficult for the air to be released by the oil.

Water Separation (Demulsification)

An emulsion is formed when two or more liquids that do not mix easily are blended together. Two well-known examples of liquids that will not mix are the oil and vinegar in a salad dressing and oil and water. In order to get them to mix, an emulsifier must be added. Some additives, such as anti-rust additives, can serve as that emulsifier.

Water, emulsified into hydraulic oil, greatly increases the collection of dust, grit, and dirt into the hydraulic fluid. Water also decreases the ability of the oil to lubricate valves, motors, and pumps. It also promotes fluid oxidation, depletes additives, and can plug filters.

Anti-Wear Properties

Fluid power systems operate at a wide variety of pressures and pump speeds. Fluids used in high-pressure systems, 1,000 psi and higher or 1,200 rpm and higher, must have excellent anti-wear capability. The high pressures or speeds in these systems do not permit the formation of a lubricating film to protect contact surfaces. Although additives exist that can increase a fluid's ability to reduce wear, and which can be added into the fluid at the end-user level, it is far better to use a balanced fluid designed for high-pressure applications.

Types of Hydraulic (Fluid Power) System Fluids

PETROLEUM

Petroleum-based hydraulic fluids are by far the most common. Their use, however, needs to be weighed against several factors.

- They can only be used where there is no danger of fire.
- They can only be used where there is no chance of the fluid contaminating other fluids, food-stuffs, and other products.
- They can only be used in systems that do not operate in a wide range of temperatures.
- They can only be used where there is no danger of environmental impact.

For the vast majority of mobile applications, all of these conditions are easily met.

FIRE-RESISTANT FLUIDS

Sometimes mobile equipment is used in environments where fire hazards are high or where environmental pollution is a major concern. These are applications where water-based (also called aqueous) fluids are advantageous. These fluids consist of water-glycols or water-oil fluids. These water-oil fluids must, by their nature, include additives such as emulsifiers, stabilizers, and additives. These fluids have low lubricity characteristics and therefore the manufacturer of the system or the manufacturer of the system components should be consulted before using these fluids in any system.

Use of these fluids should be limited to systems and components that are specifically designed to use these fluids. Many components are designed to use petroleum-based fluids specifically and can be severely damaged or suffer extremely short life cycles if fire-resistant fluids are used.

Water-Glycol Fluids

Water-glycol fluids are among the most common fire-resistant hydraulic fluids. Glycol is a glycerin/alcohol emulsion that in most forms is toxic. Ethylene glycol (often used in automotive antifreeze) and diethylene glycol are common forms of hydraulic fluid glycol. Propylene glycol is becoming a more common base for nonflammable hydraulic fluid because of low

toxicity and its ability to biodegrade. All have a very low flammability.

Water-glycol fluids usually contain the full range of additives to enhance performance. These include anti-wear, anti-foam, anti-rust, and anti-corrosion inhibitors. Fluid temperature on these systems is also very critical. Evaporation and fluid deterioration can begin at temperatures over 120°F (50°C). At temperatures below 32°F (0°C), the water-glycol blend can begin to de-emulsify and separate.

Viscosity, pH, and Water Hardness

While petroleum-based fluids require regular inspection and monitoring, water-glycol–based systems require much more monitoring. As water in the fluid is lost through evaporation, the fluid viscosity will increase. This increases friction and therefore heat in the system. Sluggish performance of the fluid power system, as well as increased power consumption, will result.

The level of **alkaline** versus acidity is referred to as pH **(Figure 5-11)**. The pH factor for distilled water is 7.

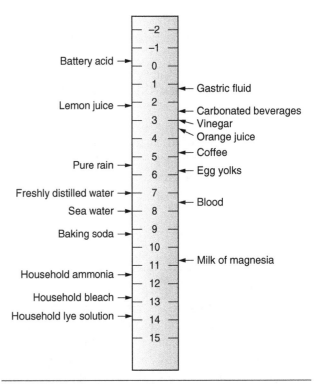

Figure 5-11 Various contaminants and deteriorating additives can cause the hydraulic fluid to form acids and bases (alkaline substances) that can damage components. The level of damage increases with the length of exposure. Typically, when the oil tends toward the acidic, it will attack the metallic components of the fluid power system. When it leans toward the alkaline, it will attack organic components such as some rubber seals and even the integrity of the hydraulic oil itself.

As the number decreases, the acidity increases. As the number decreases, the alkaline increases. With water-glycol–based systems, the pH factor should be a little above 8 at all times. This means acidity is low and the components of the system are less likely to rust. The tendency to rust increases as the level of acidity increases. Water evaporation from the fluid tends to decrease the pH rendering the fluid more acidic. Although chemicals could be added to restore the proper pH, these chemicals are very volatile and dangerous to handle. Therefore, when the pH drops below 8, the hydraulic fluid in the system should simply be replaced.

When water is added to a water-glycol system, the water that is added must be distilled or soft deionized. Two of the primary substances in hard water are calcium and magnesium. These chemicals can cause the lubricant additives to come out of solution or *floc*. This will reduce the ability of the fluid to lubricate as it works. To ensure long fluid and component life, the maximum hardness of any water added to a glycol-water hydraulic fluid should not exceed five parts per million.

Water-Oil Emulsions

- Oil-in-water: This type of fluid has excellent fire resistance, but poor lubricity. Additionally, these fluids have low viscosities, and good cooling capabilities. Oil-in-water fluids consist of mostly water with small droplets of oil dispersed evenly throughout the fluid.
- Water-in-oil: In these fluids, the percentage of oil in the fluid far exceeds the amount of water. Although water content in these fluids can be as high as 40 percent, they have very good lubrication characteristics. These fluids consist of tiny droplets of water dispersed in the oil. The oil gives good lubrication and wear characteristics to the fluid, while the droplets of water improve cooling and increase fire resistance.

All water emulsion oils should be protected from repeated cycles of freezing and thawing.

Synthetic Fire-Resistant Fluids

Three types of synthetic fire-resistant fluids are manufactured: phosphate esters, chlorinated (halogenated) hydrocarbons, and synthetic bases (a mixture of these two). These fluids are synthesized from nonvolatile materials and do not contain water. Synthetic fluids also do well in high-pressure environments. Their relatively low viscosity rating does mean that their use should be restricted to systems that have a relatively consistent operating temperature. Another factor that limits their appropriateness for different systems is their high specific gravity; this limits their ability to be pushed into the pump inlet and increases the likelihood of pump cavitation. Their fire resistance is the most impressive aspect of synthetic fire retardant fluids. Many have flash points as high as 400°F (204°C). Their self-ignition point can be as high as 900°F or 480°C.

In addition to being nonflammable, synthetic fluids are inert, odorless, non-corrosive, and have a low level of toxicity. Seal compatibility is very important when using synthetic fluids. Most of the commonly used seal materials (nitrile, neoprene, etc.) employed in fluid power systems are not compatible with synthetic fluids.

Summary

The hydraulic fluid passes through every major component in a fluid power system. The quality, additives, and condition of the fluid are absolutely essential to longevity of the system. In spite of this many fluid power systems operate their entire life on their original fluid. Meanwhile, operators of this equipment are often complaining about a decrease in efficiency, capacity, or capability. Every fluid power equipment manufacturer publishes a maintenance schedule. The schedule should be adhered to as closely as is practical. All manufacturers' recommendations should be followed, including the type and characteristics of the hydraulic oil being used in the system.

Review Questions

1. Hydraulic fluids that are rated for use in environments where there are open flames are referred to as:

    A. Water-glycol fluids.   C. Water-oil emulsions.

    B. Synthetic fire resistant fluids.   D. Any of the above

2. When hydraulic oil is said to be alkaline, it means:

    A. The acid content of the oil is high.

    B. The acid content of the oil is low.

    C. The oil is not acidic at all; rather it is the opposite of acidic.

    D. Alkaline pertains to batteries, not to hydraulic oil.

3. Technician A says hydraulic oils never need to be replaced; they only require maintenance of the additives. Technician B says that hydraulic oil only needs to be replaced when a system is used in an extreme environment with high dust content or extremes of temperature. Who is correct?

    A. Technician A only

    B. Technician B only

    C. Both Technician A and Technician B

    D. Neither Technician A nor Technician B

4. Technician A says that there can never be enough additives in the hydraulic fluid; they will only enhance the performance of the equipment. Technician B says that some additives can interfere with the performance of others. Who is correct?

    A. Technician A only

    B. Technician B only

    C. Both Technician A and Technician B

    D. Neither Technician A nor Technician B

5. The property of hydraulic fluid that gives it the ability to act as a sealing agent is its:

    A. Viscosity.

    B. Surface tension.

    C. Lubricity.

    D. Sealnicity.

6. Hydraulic fluid compressibility varies greatly depending on the type of fluid and its age.

    A. True

    B. False

    C. Varies with the type of fluid but not the age.

    D. Varies with the age of the fluid but not the type.

7. Air entrainment occurs as a result of:

    A. Agitation of the oil.

    B. Oil being exposed to high temperatures.

    C. The presence of moisture in the oil.

    D. A breakdown in one of the additives.

    E. All of the above

    F. All of the above simultaneously

8. Sulfuric acid occurs in the hydraulic oil because there are trace amounts of sulfur in many hydraulic oils. The hydraulic oils that are least likely to have significant enough sulfur to form sulfuric acid are:

    A. Synthetic.

    B. Water-based glycol.

    C. Water-oil emulsions.

    D. A and B

    E. B and C

    F. A, B, and C

9. One of the advantages of using oil in a hydraulic system is that once it absorbs heat it retains that heat for a long time. This is essential so as to make sure that all lubricated parts are continuously being made in warm hydraulic fluid.

    A. The statement above is true.

    B. The statement above is false.

    C. Heat is not a factor in hydraulic systems.

    D. Only very cold fluid can effectively lubricate the components of a fluid power system.

10. Viscosity is a characteristic of hydraulic oil that describes its ability to lubricate.

    A. True

    B. False

CHAPTER

6 Piping, Tubing, and Hoses

Learning Objectives

Upon completion and review of this chapter, the student should be able to:

- Distinguish between the types of hoses and the ratings of the hoses.
- Describe the proper type of hose used with a particular job.
- Describe the proper way to install hoses.

Cautions for This Chapter

- Hydraulic pipes, tubes, and conduits are designed to hold many times the pressure to which they will be exposed. In spite of this, they should be treated as though they might fail at any time.
- Always wear gloves that are resistant to the type of oil being handled.
- Always wear a face shield or at least safety glasses when working with hydraulic fluids.
- Spilled oil can be very slick and hazardous to walk on or handle equipment on.

Key Terms

American National Standards Institute

extreme high-pressure hose

high-pressure hose

Society of Automotive Engineers

STANDARDS

Mobile equipment fluid power systems face challenges that stationary applications do not face. In addition to the heat, vibration, and foreign object exposure of normal operation, the mobile system must also face the heat, varying environments, and vibration of the operation of the vehicle on which the fluid power system is mounted. The hoses used on mobile systems will therefore usually comply with the **Society of Automotive Engineers** (SAE) J517 standard.

Mobile fluid power systems also have the issue of being repaired and maintained by operators and technicians that are truck mechanics and not experts in fluid power systems. Mistakes are often made that go far beyond poor maintenance procedures and long maintenance intervals. Several key mistakes made by these users and mechanics contribute to system failures and poor operation. Some of the most significant mistakes relate to the hoses used in the fluid power system. These mistakes include:

- Bending the hose to less than the specified minimum radius
- Twisting, pulling, kinking, crushing, or abrading the hose

- Operating the fluid power system above the maximum temperature or below the minimum temperature
- Exposing the hoses to rapid or transient pressure rises (surges) that exceed the maximum operating pressure
- Intermixing hose types and fittings, or failing to follow manufacturer assembly or repair recommendations

In addition to SAE J517, hoses used on mobile fluid power systems should be compliant with:

- **American National Standards Institute** (ANSI) Standard T2.24.1-1991 "Hydraulic Fluid Power—Systems Standard for Stationary Industrial Machinery"
- SAE J1273 "Selection, Installation, and Maintenance of Hose and Hose Assemblies"

The critical point of all this is that any adaptations, modifications, or "make-do" component replacements alter the operating characteristics of a mobile fluid power system. Such changes can decrease the efficiency or even render the system dangerous.

Hose Categories

One of the more common mistakes made by inexperienced technicians is replacing hoses with hose material that is incorrect for the application. This can result in unnecessary expense in repairs, or worse a ruptured replacement hose and an environmental issue. In spite of all the engineering and high-grade materials that go into the production of hydraulic system hoses, they do degrade. Time, environmental conditions, flexing, vibration, and improper installation all lead to their deterioration **(Figure 6-1)**.

Hydraulic hoses consist of three layers **(Figure 6-2)**. Each of these layers may themselves consist of several layers. The innermost layer is called the tube. The specific purpose of this layer is to provide a non-porous passageway for the fluid to flow. The second layer from the center out is the reinforcement layer or layers. These layers are usually made of steel wire or a high-tensile strength fabric. The last and outer layer is called the cover and is chosen for its ability to resist abrasion, oil, corrosive materials, heat, sunlight, cold, and other environmental conditions.

EXTREME HIGH-PRESSURE HOSES

Extreme high-pressure hoses are capable of handling pressures as high as 6,000 psi. The tube is usually made of a synthetic rubber such as nitrile. This tube is then wrapped in up to six layers of spiraled steel wire.

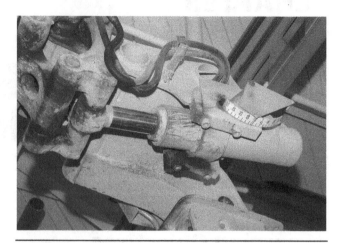

Figure 6-1 The proper installation of the appropriate type of hose can result in long-lasting, problem-free operation for the equipment. When hoses are installed, they must be of sufficient length to allow for the free articulation of the equipment. The hoses should also be bracketed to prevent them from rubbing on components as the equipment moves. Eventually normal equipment articulation will result in deterioration of the hose. When this occurs, the hose should be replaced immediately since it poses a threat to the operation of the equipment and thus is a safety hazard.

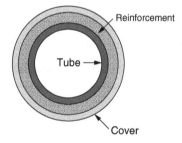

Figure 6-2 High-pressure hydraulic hoses are made up of multiple layers. The innermost layer is called a tube and is designed to be impervious to hydraulic oil. The material it is made up of is extremely dense and will not allow the hydraulic oil to pass through it. The next layer is generally a softer layer that is designed to add radial strength to the hose. Without this layer, called the reinforcing layer, the hose would burst when high pressures are applied to it. This layer is usually made from a metal or fabric similar to steel or nylon or Kevlar. The outermost layer is designed to protect both the reinforcing layer and the tube from abrasion and deterioration from outside materials. This outer covering is designed to be resistant to water, dirt, acid, caustic substances, and other materials that could damage the hose. In spite of what the outer covering is made of it is not impervious to these substances, only resistant to them.

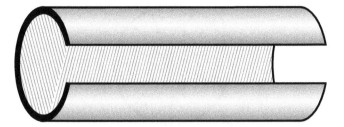

Figure 6-3 High-pressure and extreme high-pressure lines are usually reinforced with thick, densely wound fiber or steel wire. The spiral windings are usually arranged in opposing layers. The fiber or wire spiral is first wound in one direction around the hose at about a 30- to 45-degree angle, and then wound again in the second layer, this time with the angle offset by 30 to 45 degrees in the opposite direction. This makes for an extremely strong hose with a very high burst pressure; however, it does make the hose very stiff and not suitable for high angles of articulation.

Each layer of steel wire is spiraled in the opposite direction from the last **(Figure 6-3)**. This significantly increases the burst pressure of the hose. The outer layer is usually an abrasion-resistant synthetic rubber such as neoprene. Although these hoses are capable of withstanding extremely high pressures, they are stiff and difficult to bend. The spiral wire layers contribute great strength but limit how far the hose can be bent.

HIGH-PRESSURE HOSES

High-pressure hoses also feature a tube made of a synthetic rubber such as nitrile. There are usually two layers of wire wrap. These wires are braided rather than spiral wrapped **(Figure 6-4)**. This provides better

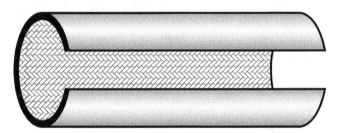

Figure 6-4 Typical high-pressure hoses, instead of using spiral wound fabric or steel, use a high-tensile-strength-rated fabric. This allows for much greater flexibility and much higher angles of articulation. The braid not only provides added strength to the hose but also reduces the likelihood that the hose will kink during high angles of movement. The brain is much thinner than the fiber or wire spiral and therefore is not capable of providing as much strength. Therefore, this design cannot be used in systems that operate at extreme high pressure.

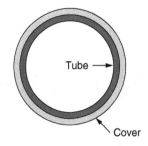

Figure 6-5 Low-pressure hoses are much simpler in design than their high-pressure counterparts. These are most often used as return lines in the typical fluid power system. Typically, a low-pressure line will consist of only the tube and a protective cover. Even the protective cover is often of lower quality than the cover on high-pressure lines.

flexibility than is found in extreme high-pressure hoses. New materials such as polyester are used in some high-pressure hoses rather than steel wire to further improve flexibility. The cover is usually a synthetic rubber such as neoprene of nitrile. Although high-pressure hoses cannot withstand the pressures that are handled by the extreme high-pressure hoses, they are more flexible.

MEDIUM-PRESSURE HOSES

Like the preceding hoses, the tube of a medium-pressure hose is an oil-resistance synthetic rubber such as nitrile. The reinforcement is typically a braided layer of steel wire or two layers of braided fabric. Like the other hoses, the cover is a synthetic rubber designed to protect the braided reinforcement and the tube.

LOW-PRESSURE HOSES

Again the tube is a synthetic rubber such as nitrile. The reinforcement is minimal, usually one fiber braid **(Figure 6-5)**. Low-pressure hoses are the most flexible and are even easily kinked if bent to a tight radius. The cover can be synthetic rubber or even synthetic rubber impregnated fabric.

Metal Pipes (Tubing)

In applications where flexibility is not required and where vibration and bending are minimal, metal tubing is used as a cheaper and more robust alternative to hoses **(Figure 6-6)**. Care must be taken in choosing replacement pipe for a system. The type of metal, the diameter, the tensile strength, and the outside diameter of the tubing can not only affect the flow through the pipe but can also affect the strength of the pipe.

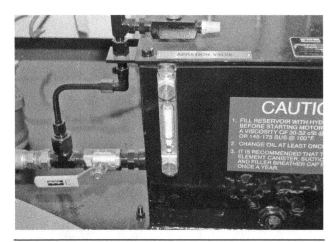

Figure 6-6 At points in a system where there is no articulation, it is common for hard steel lines to be used. Steel lines are usually cheaper and in many cases sturdier than "rubber hoses." Where steel lines are used must be considered carefully because they are far more susceptible to damage from vibration and sudden impact than is a hose. Although it is probably obvious that a solid steel line cannot be used at any point where articulation is required.

$$\text{Pressure} = \frac{(2 \times \text{wall thickness} \times \text{tensile strength})}{\text{outside diameter}}$$

Figure 6-7 The shape of both hoses and steel lines is basically an extended cylinder. As these hoses and lines carry high pressures throughout the fluid power system, pressure is exerted in all directions equally. The equalized pressure is also being applied to the surface of the inner wall of the tube in a manner that is basically perpendicular to the wall of the tube. This distributes the force of the high-pressure fluid evenly across the surface and depth of the tube. The result is that the high-pressure fluid is contained, and the burst pressure of the hose is high. Depending on the reinforcement, the burst pressure can be several thousands of psi.

OPERATING PRESSURES OF PIPES AND TUBING

The amount of pressure that can be safely accommodated by a metal pipe is calculated using Barlow's formula. The tensile strength of the material must be obtained from the manufacturer of the pipe.

Pressure = (2 × wall thickness × tensile strength) ÷ outside diameter

Pressure = (2 × 0.035 × 55,000) ÷ 0.5

Pressure = (0.070 × 55,000) ÷ 0.5

Pressure = 3,850 ÷ 0.5

Pressure = 7,700 psi

Tubing strength is, in part, a result of its cylindrical shape. The force of the hydraulic fluid is distributed evenly across all of the tubing's surfaces **(Figure 6-7)**. The result is the burst pressure of this pipe: 7,700 psi. When any metal pipe is constantly worked near its burst pressure, it will fail quickly. Therefore it is always recommended that a safety factor of at least 6 or 8 be used. This means that the maximum operating pressure to which the pipe will be exposed be 1/6 or 1/8 of the burst pressure. With a safety factor 8, the maximum operating pressure for the preceding example is 7,700 divided by 8, or 962.3 psi.

Hose and steel tubing manufacturers provide tables for use in designing fluid power systems. Although using these tables is not an everyday occurrence for the typical technician, they are an essential reference for technicians working on obsolete equipment or equipment produced by a manufacturer that is no longer in business. **Table 6-1** shows the strength characteristics of carbon steel tubing. **Table 6-2** shows the strength characteristics of copper tubing. **Table 6-3** shows the strength characteristics of stainless steel. These tubing charts are intended to be educational examples. Similar charts supplied by the tubing manufacturer should be used when actually choosing tubing.

TABLE 6-1: CARBON STEEL TUBING DATA (WITH A TENSILE STRENGTH OF 55,000 PSI)						
Tube O.D.	Wall Thickness	Tube I.D.	Inside Area	Burst psi	Working psi @ 6*	Working psi @ 8*
1/8	0.028	0.069	0.0037	24,640	4,107	3,080
	0.032	0.061	0.0029	28,160	4,693	3,520
	0.035	0.055	0.0024	30,800	5,133	3,850
3/16	0.032	0.1235	0.0120	18,773	3,130	2,347
	0.035	0.1175	0.0108	20,533	3,422	2,567

	TABLE 6-1: (*continued*)					
Tube O.D.	Wall Thickness	Tube I.D.	Inside Area	Burst psi	Working psi @ 6*	Working psi @ 8*
1/4	0.035	0.180	0.0254	15,400	2,567	1,925
	0.042	0.166	0.0216	18,480	3,080	2,310
	0.049	0.152	0.0181	21,560	3,593	2,695
	0.058	0.134	0.0141	25,520	4,253	3,190
	0.065	0.120	0.0113	28,600	4,767	3,575
5/16	0.035	0.2425	0.0462	12,320	2,053	1,540
	0.042	0.2285	0.0410	14,784	2,464	1,848
	0.049	0.2145	0.0361	17,248	2,875	2,156
	0.058	0.1965	0.0303	20,416	3,403	2,552
	0.065	0.1825	0.0262	22,880	3,813	2,860
3/8	0.035	0.305	0.0730	10,267	1,711	1,283
	0.042	0.291	0.0665	12,320	2,053	1,540
	0.049	0.277	0.0602	14,373	2,396	1,797
	0.058	0.259	0.0527	17,013	2,835	2,127
	0.065	0.245	0.0471	19,067	3,178	2,383
1/2	0.035	0.430	0.1452	7,700	1,283	963
	0.042	0.416	0.1359	9,240	1,540	1,155
	0.049	0.402	0.1269	10,780	1,797	1,348
	0.058	0.384	0.1158	12,760	2,127	1,595
	0.065	0.370	0.1075	14,300	2,383	1,788
	0.072	0.356	0.0995	15,840	2,640	1,980
	0.083	0.334	0.0876	18,260	3,043	2,283
5/8	0.035	0.555	0.2418	6,160	1,027	770
	0.042	0.541	0.2298	7,392	1,232	924
	0.049	0.527	0.2180	8,624	1,437	1,078
	0.058	0.509	0.2034	10,208	1,701	1,276
	0.065	0.495	0.1923	11,440	1,907	1,430
	0.072	0.481	0.1816	12,672	2,112	1,584
	0.083	0.459	0.1654	14,608	2,435	1,826
	0.095	0.435	0.1485	16,720	2,787	2,090
3/4	0.049	0.652	0.3337	7,187	1,198	898
	0.058	0.634	0.3155	8,507	1,418	1,063
	0.065	0.620	0.3018	9,533	1,589	1,192
	0.072	0.606	0.2813	10,560	1,760	1,320
	0.083	0.584	0.2677	12,173	2,029	1,522
	0.095	0.560	0.2462	13,933	2,322	1,742
	0.109	0.532	0.2222	15,987	2,664	1,998
7/8	0.049	0.777	0.4739	6,160	1,027	770
	0.058	0.759	0.4522	7,291	1,215	911
	0.065	0.745	0.4357	8,171	1,362	1,021
	0.072	0.731	0.4195	9,051	1,509	1,131
	0.083	0.709	0.3946	10,434	1,739	1,304
	0.095	0.685	0.3683	11,943	1,990	1,493
	0.109	0.657	0.3388	13,703	2,284	1,713
1	0.049	0.902	0.6387	5,390	898	674
	0.058	0.884	0.6134	6,380	1,063	798
	0.065	0.870	0.5942	7,150	1,192	894
	0.072	0.856	0.5752	7,920	1,320	990
	0.083	0.834	0.5460	9,130	1,522	1,141
	0.095	0.810	0.5150	10,450	1,742	1,306
	0.109	0.782	0.4801	11,990	1,998	1,500
	0.120	0.760	0.4534	13,200	2,200	1,650
1 1/4	0.490	1.152	1.0418	4,312	719	539
	0.058	1.134	1.0095	5,104	851	638

(*continued*)

TABLE 6-1: CARBON STEEL TUBING DATA (WITH A TENSILE STRENGTH OF 55,000 PSI) (continued)

Tube O.D.	Wall Thickness	Tube I.D.	Inside Area	Burst psi	Working psi @ 6*	Working psi @ 8*
	0.065	1.120	0.9847	5,720	953	715
	0.072	1.106	0.9602	6,336	1,056	792
	0.083	1.084	0.9224	7,304	1,217	913
	0.095	1.060	0.8820	8,360	1,393	1,045
	0.109	1.032	0.8360	9,592	1,600	1,200
	0.120	1.010	0.8008	10,560	1,760	1,320
1 1/2	0.065	1.370	1.4734	4,767	794	596
	0.072	1.356	1.4434	5,280	880	660
	0.083	1.334	1.3970	6,087	1,014	761
	0.095	1.310	1.3471	6,967	1,161	871
	0.109	1.282	1.2902	7,993	1,332	1,000
	0.120	1.260	1.2463	8,800	1,467	1,100
1 3/4	0.065	1.620	2.0602	4,086	681	511
	0.072	1.606	2.0247	4,526	754	566
	0.083	1.584	1.9696	5,217	870	652
	0.095	1.560	1.9104	5,971	995	746
	0.109	1.532	1.8424	6,851	1,142	856
	0.120	1.510	1.7899	7,543	1,257	943
	0.134	1.482	1.7241	8,423	1,404	1,053
2	0.065	1.870	2.7451	3,575	596	447
	0.072	1.856	2.7041	3,960	660	495
	0.083	1.834	2.6404	4,565	761	571
	0.095	1.810	2.5717	5,225	871	653
	0.109	1.782	2.4928	5,995	1,000	749
	0.120	1.760	2.4316	6,600	1,100	825
	0.134	1.732	2.3549	7,370	1,228	921

Note: These specs are intended to be used as examples only.
PSI 6* = a safety factor of 1/6 burst pressure
PSI 8* = a safety factor of 1/8 burst pressure

Copper

TABLE 6-2: COPPER TUBING DATA (WITH A TENSILE STRENGTH OF 32,000 PSI)

Tube O.D.	Wall Thickness	Tube I.D.	Inside Area	Burst psi	Working psi @ 6*	Working psi @ 8*
1/4	0.030	0.190	0.0283	7,680	1,280	960
	0.049	0.152	0.0181	12,544	2,090	1,568
5/16	0.032	0.249	0.0485	6,554	1,092	819
	0.049	0.215	0.0361	10,035	1,673	1,254
3/8	0.032	0.311	0.0759	5,461	910	683
	0.058	0.259	0.0527	9,899	1,650	1,237
	0.072	0.231	0.0419	12,288	2,048	1,536
1/2	0.032	0.436	0.1492	4,096	683	512
	0.049	0.402	0.1269	6,272	1,045	784
	0.058	0.384	0.1158	7,424	1,237	928
	0.072	0.356	0.0995	5,376	896	672
5/8	0.035	0.555	0.2418	3,584	597	448
	0.049	0.527	0.2180	5,018	836	627
	0.065	0.495	0.1923	6,656	1,109	832

TABLE 6-2: (*continued*)						
Tube O.D.	Wall Thickness	Tube I.D.	Inside Area	Burst psi	Working psi @ 6*	Working psi @ 8*
3/4	0.035	0.680	0.3630	2,987	498	373
	0.049	0.652	0.3371	4,181	697	523
	0.065	0.620	0.3018	5,547	924	693
7/8	0.045	0.785	0.4837	3,291	549	411
	0.065	0.745	0.4357	4,754	792	594
1	0.065	0.870	0.5942	4,160	693	520
1 1/8	0.050	1.025	0.8247	2,844	474	356
1 1/4	0.083	1.084	0.9224	4,250	708	531
1 3/4	0.055	1.265	1.2562	2,560	427	320

Note: These specs are intended to be used as examples only.
PSI 6* = a safety factor of 1/6 burst pressure
PSI 8* = a safety factor of 1/8 burst pressure

Stainless Steel

TABLE 6-3: STAINLESS STEEL TUBING DATA (WITH A TENSILE STRENGTH OF 75,000 PSI)						
Tube O.D.	Wall Thickness	Tube I.D.	Inside Area	Burst psi	Working psi @ 6*	Working psi @ 8*
1/8	0.032	0.061	0.0029	38,400	6,400	4,800
3/16	0.032	0.124	0.0120	25,600	4,267	3,200
	0.035	0.118	0.0108	28,000	4,667	3,500
1/4	0.035	0.180	0.0254	21,000	3,500	2,625
	0.049	0.152	0.0181	29,400	4,900	3,675
5/16	0.035	0.243	0.0462	16,800	2,800	1,000
	0.049	0.215	0.0361	23,520	3,920	2,940
	0.058	0.197	0.0303	27,840	4,640	3,480
3/8	0.035	0.305	0.0730	14,000	2,333	1,750
	0.049	0.277	0.0602	19,600	3,267	2,456
	0.058	0.259	0.0527	23,200	3,867	2,900
	0.065	0.245	0.0471	26,000	4,333	3,250
1/2	0.035	0.430	0.1452	10,500	1,750	1,313
	0.049	0.402	0.1269	14,700	2,450	1,842
	0.058	0.384	0.1158	17,400	2,900	2,175
	0.065	0.370	0.1075	19,500	3,250	2,438
	0.072	0.356	0.0995	21,600	3,600	2,700
	0.083	0.334	0.0876	24,900	4,150	3,113
5/8	0.049	0.527	0.2180	11,760	1,960	1,470
	0.058	0.509	0.2034	13,920	2,320	1,740
	0.065	0.495	0.1923	15,600	2,600	1,950
	0.072	0.481	0.1816	17,280	2,880	2,160
	0.083	0.459	0.1654	19,920	3,320	2,490
	0.095	0.435	0.1485	22,800	3,800	2,850
3/4	0.049	0.652	0.3337	9,800	1,633	1,225
	0.058	0.634	0.3155	11,600	1,933	1,450
	0.065	0.620	0.3018	13,000	2,167	1,625
	0.072	0.606	0.2883	14,400	2,400	1,800
	0.083	0.584	0.2677	16,600	2,767	2,075
	0.095	0.560	0.2462	19,000	3,167	2,375

(*continued*)

TABLE 6-3: STAINLESS STEEL TUBING DATA (WITH A TENSILE STRENGTH OF 75,000 PSI) (*continued*)

Tube O.D.	Wall Thickness	Tube I.D.	Inside Area	Burst psi	Working psi @ 6*	Working psi @ 8*
7/8	0.049	0.777	0.4739	8,400	1,400	1,050
	0.058	0.759	0.4522	9,943	1,657	1,243
	0.065	0.745	0.4357	11,143	1,857	1,393
	0.072	0.731	0.4195	12,343	2,057	1,543
	0.083	0.709	0.3946	14,229	2,371	1,779
	0.095	0.685	0.3683	18,153	3,025	2,269
	0.109	0.657	0.3388	18,686	3,114	2,336
1	0.049	0.902	0.6387	7,350	1,225	919
	0.058	0.884	0.6134	8,700	1,450	1,088
	0.065	0.870	0.5942	9,750	1,625	1,219
	0.072	0.856	0.5752	10,800	1,800	1,350
	0.083	0.834	0.5460	12,450	2,075	1,556
	0.095	0.810	0.5150	14,250	2,375	1,781
	0.109	0.782	0.4801	16,350	2,725	2,044
1 1/4	0.083	1.084	0.9224	9,960	1,660	1,245
	0.095	1.060	0.8820	11,400	1,900	1,425
	0.109	1.032	0.8360	13,080	2,180	1,635
	0.120	1.010	0.8008	14,400	2,400	1,800
1 1/2	0.095	1.310	1.3471	9,500	1,583	1,188
	0.109	1.282	1.2902	10,900	1,817	1,363
	0.120	1.260	1.2463	12,000	2,000	1,500
	0.134	1.232	1.1915	13,400	2,233	1,675

Note: These specs are intended to be used as examples only.
PSI 6* = a safety factor of 1/6 burst pressure
PSI 8* = a safety factor of 1/8 burst pressure

OIL FLOW CAPACITY OF PIPES AND TUBING

The amount of flow in a fluid power system is dependent on how much can flow through the components with the least flow capacity. When replacing a pipe or tube, care must be taken to ensure that the replacement does not restrict flow in the system (**Figure 6-8**). This requires proper sizing. As the size of the tubing increases at a given flow rate (gallons per minute), the velocity of the flow decreases. As the size of the tubing decreases, the velocity at a given flow rate increases. Increased velocity means increased friction. Increased friction results in increased heat in the system. **Table 6-4** is a chart illustrating flow rate when velocity is constant.

Two formulas relate to what will occur when the size of a pipe is changed in a fluid power system. The first was discussed in **Chapter 3** and relates to changes in fluid velocity if the gallons-per-minute flow rate changes. This is calculated as:

$$V = GPM \times 0.3208 \div A$$
Velocity = gallons per minute × 0.3208 ÷ A

The second formula calculates changes in the flow rate when velocity is a constant. An example of this would be replacing a 0.5-inch O.D. tube with a 0.035 wall with a 0.5-inch O.D. pipe that has a 0.042 wall. The thicker wall pipe will have a smaller cross-sectional area. The result will be a decreased flow rate for the fluid at a given velocity, or a forced increase in velocity to maintain the flow rate in gallons per minute.

$$GPM = V \times A \div 0.3208$$

OIL PRESSURE LOSS THROUGH PIPES

Oil pressure loss through pipes, tubing, and hoses is a significant issue when dealing with either very small pipes or long pipes. Generally, in mobile fluid power applications we deal with neither. Nevertheless, it is important for the service technician to recognize that changing the diameter of the hoses and pipes can adversely affect the operation of a fluid power system.

		TABLE 6-4: FLOW RATE WHEN VELOCITY IS A CONSTANT					
Tube O.D.	Wall Thickness	2 Ft/Sec GPM	4 Ft/Sec GPM	10 Ft/Sec GPM	15 Ft/Sec GPM	20 Ft/Sec GPM	30 Ft/Sec GPM
1/2	0.035	0.905	1.81	4.52	6.79	9.05	13.6
	0.042	0.847	1.63	4.23	6.35	8.47	12.7
	0.049	0.791	1.58	3.95	5.93	7.91	11.9
	0.058	0.722	1.44	3.61	5.41	7.22	10.8
	0.065	0.670	1.34	3.35	5.03	6.70	10.1
	0.072	0.620	1.24	3.10	4.65	6.20	9.30
	0.083	0.546	1.09	2.73	4.09	5.46	8.18
5/8	0.035	1.510	3.01	7.54	11.3	15.1	22.6
	0.042	1.430	2.85	7.16	10.7	14.3	21.4
	0.049	1.360	2.72	6.80	10.2	13.6	20.4
	0.058	1.270	2.54	6.34	9.51	12.7	19.0
	0.065	1.200	2.40	6.00	9.00	12.0	18.0
	0.072	1.130	2.26	5.66	8.49	11.3	17.0
	0.083	1.030	2.06	5.16	7.73	10.3	15.5
	0.095	0.926	1.85	4.63	6.95	9.26	13.9
3/4	0.049	2.08	4.17	10.4	15.6	20.8	31.2
	0.058	1.97	3.93	9.84	14.8	19.7	29.6
	0.065	1.88	3.76	9.41	14.1	18.8	28.2
	0.072	1.75	3.51	8.77	13.2	17.5	26.4
	0.083	1.67	3.34	8.35	12.5	16.7	25.0
	0.095	1.53	3.07	7.67	11.5	15.3	23.0
	0.109	1.39	2.77	6.93	10.4	13.9	20.8
7/8	0.049	2.95	5.91	14.8	22.2	29.5	44.3
	0.058	2.82	5.64	14.1	21.1	28.2	42.3
	0.065	2.72	5.43	13.6	20.4	27.2	40.7
	0.072	2.62	5.23	13.1	19.6	26.2	39.2
	0.083	2.46	4.92	12.3	18.5	24.6	36.9
	0.095	2.30	4.60	11.5	17.2	23.0	34.4
	0.109	2.11	4.22	10.6	15.8	21.1	31.7
1	0.049	3.98	7.96	19.9	29.9	39.8	59.7
	0.058	3.82	7.65	19.1	28.7	38.2	57.4
	0.065	3.70	7.41	18.5	27.8	37.0	55.6
	0.072	3.59	7.17	17.9	26.9	35.9	53.8
	0.083	3.40	6.81	17.0	25.5	34.0	51.1
	0.095	3.21	6.42	16.1	24.1	32.1	48.2
	0.109	3.00	6.00	15.0	22.4	29.9	44.9
	0.120	2.83	5.65	14.1	21.2	28.3	42.4
1 1/4	0.049	6.50	13.0	32.5	48.7	64.9	97.4
	0.058	6.29	12.6	31.5	47.2	62.9	94.4
	0.065	6.14	12.3	30.7	46.0	61.4	92.1
	0.072	6.00	12.0	30.0	44.9	59.9	89.8
	0.083	5.75	11.5	28.8	43.1	57.5	86.3
	0.095	5.50	11.0	27.5	41.2	55.0	82.5
	0.109	5.21	10.4	26.1	39.1	52.1	78.2
	0.120	5.00	10.0	25.0	37.4	50.0	74.9
1 1/2	0.065	9.19	18.4	45.9	68.9	91.9	138
	0.072	9.00	18.0	45.0	67.5	90.0	135
	0.083	8.71	17.4	43.5	65.3	87.1	131
	0.095	8.40	16.8	42.0	63.0	84.0	126
	0.109	8.04	16.1	40.2	60.3	80.4	121
	0.120	7.77	15.5	38.8	58.3	77.7	117
1 3/4	0.065	12.8	25.7	64.2	96.3	128	193
	0.072	12.6	25.2	63.1	94.7	126	189

(continued)

TABLE 6-4: FLOW RATE WHEN VELOCITY IS A CONSTANT (*continued*)							
Tube O.D.	Wall Thickness	2 Ft/Sec GPM	4 Ft/Sec GPM	10 Ft/Sec GPM	15 Ft/Sec GPM	20 Ft/Sec GPM	30 Ft/Sec GPM
	0.083	12.3	24.6	61.4	92.1	123	184
	0.095	11.9	23.8	59.6	89.3	119	179
	0.109	11.5	23.0	57.4	86.1	115	172
	0.120	11.2	22.3	55.8	83.7	112	167
	0.134	10.7	21.5	53.7	80.6	107	161
2	0.065	17.1	34.2	85.6	128	171	257
	0.072	16.9	33.7	84.3	126	169	253
	0.083	16.5	32.9	82.3	123	165	247
	0.095	16.0	32.1	80.2	120	160	240
	0.109	15.5	31.1	77.7	117	155	233
	0.120	15.2	30.3	75.8	114	152	227
	0.134	14.7	29.4	73.4	110	147	220

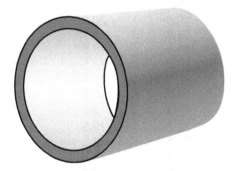

Figure 6-8 The inner layer of the hose, referred to as the tube, is made of the material that enhances flow through the hose. Though inherently weak in its structure, the smooth surface of the inner tube of the hose allows for low friction, low turbulence, and low heat buildup as the hydraulic fluid flows through the hose. In some cases, after years of use, and after years of abuse from the wrong types of oils or foreign materials being introduced into the oils, this inner tube can collapse. When it collapses, it usually rips, which allows for a hydraulic oil leak through the side wall of the hose. More subtle is the fact that the collapse of the inner tube forms a restriction to the flow of the fluid. This is often difficult to diagnose, but by using an infrared heat gun it is sometimes noticeable because of the temperature change that occurs downstream of the point of the restriction.

Figure 6-9 Every hydraulic hose is marked with information from the manufacturer. This information usually includes the name of the manufacturer, any unique features of the hose, industry standards to which the hose is built, and burst pressures. This hose is a no-skive hose. This means that when the hose was built, there was no modification, shaving, or cutting of the end of the hose to install the fittings. The numbers 301-16 WP indicate the manufacturer series, size, and type. Crucial for the technician to double-check is the rated burst pressures of the hose. This hose has a burst pressure of 13.8 megapascals. A megapascal is roughly equivalent to 150 psi. Notice that this hose is also marked in psi. The MSHA indicates that the hose has been designed and built in accordance with standards established by the Mine Safety and Health Administration.

Summary

Great care should be taken when replacing hydraulic hoses and lines. Using the wrong type of hose or line can lead to catastrophic failure, which can lead to damaged machinery, injury, and even death. Whenever possible, use parts provided by the equipment manufacturer or recommended by a reputable hydraulic supply firm. Hydraulic hoses and many metal lines are marked with identifying labels describing the characteristics of the hose. Be sure that replacement hoses and lines bear the same identifying marks as the original (**Figure 6-9**).

Review Questions

1. Hydraulic hoses are specifically designed to allow for maximum flexibility under all operating conditions, at all temperatures, and at all pressures.

   A. The above statement is true.

   B. The above statement is false.

   C. All hydraulic hoses are built to the same standard. This is to ensure that the replacement hose will always match its predecessor.

   D. There are no standards for hydraulic hoses, each manufacturer is free to design hoses to their own specifications.

2. In an extreme high-pressure hose, the reinforcing material is:

   A. Spiral wound Kevlar.

   B. Spiral wound steel wire.

   C. Both A and B

   D. Neither A nor B

3. Technician A says that when the correct type of hose is not available, any hydraulic hose can be used as a short-term repair. Technician B says it is okay to replace a hose of a higher pressure rating even though the hose may be less flexible. Who is correct?

   A. Technician A

   B. Technician B

   C. Both Technician A and Technician B

   D. Neither Technician A nor Technician B

4. The braided fabric reinforcement used in some hydraulic hoses is a characteristic of:

   A. Extreme high-pressure hoses.

   B. High-pressure hoses.

   C. Medium-pressure hoses.

   D. Return lines.

5. Which of the following metals should only be used in low-pressure and return lines?

   A. Carbon steel

   B. Stainless steel

   C. Aluminum alloy

   D. Copper

6. A section of carbon steel tubing has an outside diameter of 7/8 of an inch. The wall thickness is 0.095. Assuming we are using a safety factor of six, what is the maximum working pressure the system should have at the point where this section of tubing is located?

   A. 1,990

   B. 2,284

   C. 1,493

   D. This could not be found on the chart.

7. In the preceding question, a safety factor of six was mentioned. This means:

   A. The maximum safe operating pressure is 6 psi below the burst pressure.

   B. Multiply the burst pressure times six to calculate the maximum working pressure.

   C. The maximum working pressure is 1/6 of the burst pressure.

   D. Six inches is the minimum radius to which this type of model piping can be safely bent.

8. What is the maximum recommended working pressure of 1-inch copper tubing, assuming a safety factor of 8?

   A. 0.065

   B. 0.5942

   C. 693

   D. 520

9. According to the flow rate velocity chart found earlier in this chapter, what would be the flow rate through a 1-inch tube if the wall thickness was 0.120 and the velocity was 10 feet per second?

    A. 14.1 gallons per minute   C. 19.9 gallons per minute

    B. 21.2 gallons per minute   D. 39.8 gallons per minute

10. Technician A says there will always be pressure loss as hydraulic oil flows through pipes, tubing, and hoses. Technician B says that hydraulic pipes, tubing, and hoses are specifically designed so there is no pressure drop as the hydraulic oil flows through the system. Who is correct?

    A. Technician A   C. Both Technician A and Technician B

    B. Technician B   D. Neither Technician A nor Technician B

CHAPTER

7 Couplers and Fittings

Learning Objectives

Upon completion and review of this chapter, the student should be able to:

- Distinguish between the different types of fittings and select a fitting that is appropriate to the pressure in the system.
- Be able to properly connect and torque fluid power system settings.
- Be able to predict a pressure loss as a result of flow through an angular fitting.

Cautions for This Chapter

- Proper torque is essential when tightening fittings onto hoses and lines. One of the most common causes of damage to hydraulic lines and hoses is overtorquing.
- Always make sure the proper fittings are used when joining hoses or lines together. Mismatching the fittings can lead to system failure and personal injury.
- Overtorquing fittings can stretch the threads and provided a spiral path for leaks on many types of fittings.
- Always wear gloves that are resistant to the type of oil being handled.
- Always wear a face shield or at least safety glasses when working with hydraulic fluids.
- Spilled oil can be very slick and hazardous to walk on or handle equipment on.

Key Terms

American National Standards Institute (ANSI)

flats

International Standards Organization (ISO)

Joint Industrial Council (JIC)

Society of Automotive Engineers (SAE)

INTRODUCTION

Fittings in fluid power systems come in several varieties. The type of fitting a system employs has as much to do with the preferences of the engineering group designing the system as with the characteristics of the system being designed. Like a lot of other things, the use of coupler types and fitting types are at the discretion and preference of the engineer and designer of the system, as well as the technician maintaining the system. Some of the fitting designs have been in used both successfully and unsuccessfully for decades. Others are relatively new and have come into being because of the availability of new materials.

In this chapter, we will explore various types of couplers and fittings and discuss some of the advantages and disadvantages of each.

Familiarity with the various couplers and fittings can save the service technician a great deal of grief in dealing with leaks and poorly fitting hoses. Although they are distinctly different from one another at first glance, some of the hydraulic unions (more commonly referred to as couplers) and settings do look alike.

THREADED SEAL METAL-TO-METAL TAPERED FITTING

One of the oldest fittings used in hydraulic systems is the threaded seal metal-to-metal tapered union (**Figure 7-1**). Although when properly fitted this union can create an effective proof seal, fitting it properly can be a challenge for the less experienced service technician. The primary advantage of this type of fitting is simply cost. Simple hydraulic systems that do simple tasks for low prices will use this type of hose fitting merely as a way of reducing cost.

This type of fitting comes with two primary concerns. The first would be a common mistake made by the less experienced technician. In this type of connection, the thread-to-thread mating surfaces are the only source for sealing. If this union is overtorqued or overtightened, it can cause the threads to be distorted or stretched. This distortion can prevent a good seal, which can lead to leaks. Also, over-tightening can weaken the threads and even weaken the union itself, resulting in the failure of the fitting and thus a catastrophic leak. A secondary mistake related to this type of union is the use of improper sealants in an attempt to repair the leaks caused by the stretched threads. Overtorqued fittings should simply be replaced.

The second issue with this type of fitting is the fact that the male end often does not rotate on the hose. This can make it extremely awkward and even impossible to get a good connection without weakening the hose if both female ends that the hose is to screw into cannot be rotated on the hose. Equipment

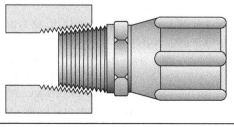

Figure 7-1 Threaded seal metal-to-metal tapered fitting.

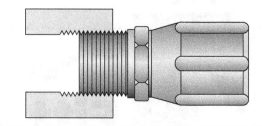

Figure 7-2 Threaded seal metal-to-metal straight fitting.

designed in this manner should be updated or upgraded instead of simply having the hose replaced.

THREADED SEAL METAL-TO-METAL STRAIGHT FITTING

Threaded seal metal-to-metal straight fittings have all the problems of the threaded seal metal-to-metal tapered fitting, plus a couple of others (**Figure 7-2**). Both when installing and when removing the threaded seal metal-to-metal straight fitting, there is a brief time when the fitting is held in place by a single thread. If the hose is allowed to put stress on the single thread during disassembly, the thread will stretch and/or distort. This distortion can occur not only on the male hose of the union but also on the stationary female end of the union. When the new hose is installed, or even when the old hose is reinstalled, getting the first thread lined up properly can be difficult. Even if it does line up properly, it is distorted, stretched, and as the hose is screwed into position, the distortion on the final thread can be tapped into the rest of the threads. The end result is that both the male and female in the union are distorted, and this can affect the quality of the seal. Again, the technician may make a further mistake of attempting to repair the resulting leak with the use of improper sealants.

Just as in the threaded seal metal-to-metal tapered fitting, the only real advantage of this type of union is low cost.

DRY SEAL TO 30-DEGREE CONE SEAT FITTING

The two previous styles of fittings depended exclusively on the threads to make a seal. As was discussed, this can work fairly well as long as the threads are in good condition and especially if the threads are not damaged, distorted, or stretched due to being overtorqued. The dry seal to 30-degree cone seat (**Figure 7-3**) uses the threads only to hold a 30-degree angle sealing junction firmly in place. As a result, if

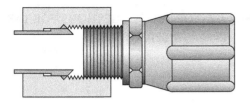

Figure 7-3 Dry seal to 30-degree cone seat fitting.

the threads have been distorted or stretched, the only impact on the integrity of the seal is the ability of the damaged threads to hold a seal firmly in place.

In the thread of the male end of the hose there is a 30-degree downward tapered semimachined surface. Inserted into the female end is a reverse-tapered machined ferule. When the male fitting is attached to the female fitting, the two angles mate and form a sealing surface. Because the angle surfaces are made of a relatively soft material, its surfaces conform to one another and form a good seal. Because the seal is not dependent on the treads of the fitting, leaks caused by stresses on the hose as the machinery operates are far less likely.

JOINT INDUSTRIAL COUNCIL (JIC) FLARE SEAL TO 37-DEGREE CONE SEAT FITTING

This type of fitting or junction features a crushable flare seal on the inside of the female end of the connection **(Figure 7-4)**. The female side of the connection is actually a flare nut in which the flared tubing junction is seated. The male end features a 37-degree cone taper that presses firmly into the crushable seal of the flare. The end result is a virtually leak-free seal even with a fairly low torque applied to the coupling. Proper torque with this type of fitting is necessary to ensure that the threads are not damaged. Furthermore, it is essential that the 37-degree cone taper surface be free of contamination, particularly hard particles such as sand and dirt, when the initial mating of the two surfaces happens. This ensures that a good seal will be made when the connection is disassembled and then reassembled.

The **Joint Industrial Council (JIC)** flare seal is particularly well-suited for high-pressure applications

TABLE 7-1: 37-DEGREE JIC FLARE SEAL TIGHTENING

Flats from Finger Tight	Size	Lb-Ft Min/Max
–	–02	6–7
–	–03	8–9
2	–04	11–12
2	–05	14–15
1 1/2	–06	18–20
1 1/2	–08	36–39
1 1/2	–10	57–63
1 1/4	–12	79–88
1	–14	94–103
1	–16	108–113
1	–20	127–133
1	–24	158–167
1	–32	245–258

where flexibility and light weight are essential. The sleeve absorbs vibrations, and because the seal is metal to metal, they can be reliably connected and reconnected multiple times. The requirement that the tubing end be flared does not make the 37-degree JIC flare suitable for use in systems that use thick wall tubing. These fittings are available in sizes from 1/8 inch up to 2 inches **(Table 7-1)**. Because torquing is critical, make sure that these recommendations, or similar recommendations from the manufacturing source, are followed.

SAE FLARE SEAL TO 45-DEGREE CONE SEAT FITTING

The SAE flare seal to 45-degree cone seat **(Figure 7-5)** is similar in most respects to the 37-degree cone seat of the JIC. The 45-degree cone seat, however, is used primarily in hydraulic brake systems and power steering systems. As a result, this type of fitting is only commonly available in two sizes: the 3/8-inch I.D. and 3/4-inch I.D. sizes. Some smaller hydraulic equipment intended for mobile application and meant to be compatible with components more readily

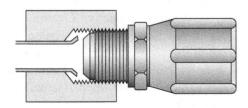

Figure 7-4 JIC flare seal to 37-degree cone seat fitting.

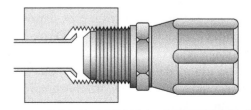

Figure 7-5 SAE Flare seal to 45-degree cone seat fitting.

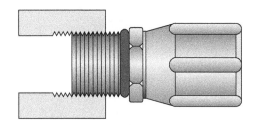

Figure 7-6 O-ring seat fitting.

available than standard hydraulic fittings utilize the SAE flare seal and the 45-degree cone seat.

O-RING SEAT FITTING

The standard O-ring seat fitting **(Figure 7-6)** uses exactly the same threads as the 37-degree JIC fitting. In the O-ring seat fitting, however, two parallel metal surfaces are mated with a rubber-like O-ring between the surfaces. The O-ring is almost never natural rubber, but in most cases is made of nitrile or some other highly resilient compound. This type of seal offers one of the best leak-free connections in a fluid power system. The rubber compensates for any nicks, small gouges, or other imperfections in the mating surfaces. It is highly recommended for medium- and high-pressure systems. The fitting can be reused many times as long as the O-ring is replaced each time the fitting is disconnected and reconnected. Minimal damage to the mating surfaces takes place because there is no crushing of metal, flares, or the mating surfaces themselves to form a seal.

As with the 37-degree JIC fitting, proper torque is critical when joining or rejoining the fittings together. Most technicians will attempt to guesstimate the torque as they tighten the fitting. Tests have shown that in almost all cases this leads to overtightening of the fitting **(see Table 7-2)**.

SPLIT FLANGE O-RING SEAL FITTING

The split flange design **(Figure 7-7)** is used worldwide for connecting pumps and hydraulic motors in fluid power systems. The hose or line fitting has a thick lip near the end of the fitting. A narrow protrusion provides a mounting point for an O ring or a quad ring. A flange bracket is then placed on either side of the line on top of the thick lip. Two bolts hold each of the flange brackets in place, firmly compressing the O-ring against the mounting of the hydraulic pump or motor. This makes for a firm, resilient, and dependable assembly. This connection leaves no spiral leak paths

TABLE 7-2: TORQUE SUGGESTIONS FOR O-RING SEAT FITTINGS	
Size	**Lb-Ft Min/Max**
−02	6–7
−03	8–10
−04	13–15
−05	17–21
−06	22–25
−08	40–43
−10	43–57
−12	68–75
−14	90–99
−16	112–123
−20	146–200
−24	154–215
−32	218–290

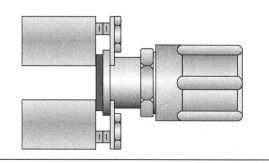

Figure 7-7 Split flange O-ring seal fitting.

such as those found in most threaded fittings. There is also no significant possibility of gasket misalignment and no need for sealing compounds.

As always, it is important to make sure that the surface of the port to which the line is being attached and the mounting surface for the O-ring are clean and free of burrs, nicks, dirt, or other foreign materials. This type of fitting can be reused multiple times and only requires a new O-ring each time the fitting is disconnected and reconnected. As with almost all fittings, it is necessary to make sure that the bolts attaching the flanges are torqued to the proper spec. The split flange design may make it even more critical that this fitting be torqued properly. Overtorquing the flanges can result in the O-ring being literally squeezed out of its proper position, creating a pathway for leaks. Refer to the torquing chart **(Table 7-3)** or the manufacturer's torquing chart when assembling a split flange O-ring fitting.

Note that in the preceding chart, reference is made to code 61 versus code 62 fittings. Code 61 and code

TABLE 7-3: TORQUE SUGGESTIONS FOR SPLIT FLANGE O-RING SEALS				
Code 61			Code 62	
Bolts	Lb-Ft	Size	Bolts	Lb-Ft
3/8–16	21–29	–12	3/8–16	26–34
3/8–16	26–35	–16	7/16–14	26–50
7/16–14	36–46	–20	1/2–13	63–75
1/2–13	46–58	–24	5/8–11	117–153
1/2–13	56–66	–32		

TABLE 7-4: TORQUE SUGGESTIONS FOR O-RING FACE SEAL	
–04	10–12
–06	18–20
–08	32–35
–10	46–50
–12	65–70
–16	92–100
–20	125–140
–24	150–165

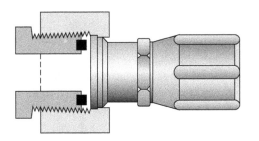

Figure 7-8 O-ring face seal fitting.

62 are related to a **Society of Automotive Engineers (SAE)** standard called SAE J 1518. The standard states that fittings meeting the code 61 standard will be viable up to 5,000 psi, and those meeting the code 62 standard will be viable up to pressures of 6,000 psi.

O-RING FACE SEAL FITTING

OFS or O-ring face seal fittings **(Figure 7-8)** have become very popular in the last couple of decades. One of the primary features is the fact that they can be significantly overtorqued without damaging the O-ring or the fitting, thereby yielding an effective seal. The O-ring sits over the end of the fitting in much the same way it does in the split lands design. However, there is a sleeve around the outer circumference of the O-ring that keeps the O-ring in position and prevents overtorquing from crushing it. When torqued down, the union makes a metal-to-metal contact, properly compressing the O-ring within the groove in which it is mounted.

A unique feature of the O-ring face seal is that there is a much higher torque spec associated with these fittings for a given size **(Table 7-4)**.

QUICK DISCONNECT FITTING

Quick disconnect fittings **(Figure 7-9)** are used where tools, accessories, or test equipment are to be installed. Several different types and sizes are available, therefore great care should be taken to make sure the

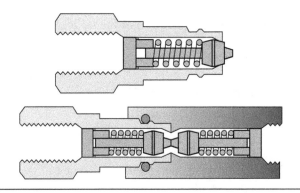

Figure 7-9 Quick disconnect fitting.

fitting is replaced with the correct part. Failure to do so could render the equipment unusable in the field.

OIL FLOW THROUGH ORIFICES

By their very nature, all fittings create some restriction in oil flow. In some cases, the fitting has been specifically designed to form a restriction to limit the flow at some point in the fluid power system **(Figure 7-10)**. Several reasons may exist to install a restricting orifice or other device in a fluid power system. This could be to increase line pressure, to decrease the flow through a line, or to increase the fluid velocity in the line.

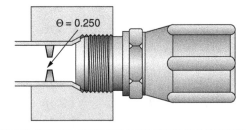

Figure 7-10 Any restriction in a hydraulic system has an impact on the flow rate of fluid through the system. The person designing, repairing, or troubleshooting a system needs to remember that every fitting is inherently a flow-restriction orifice.

The formula for calculating the effective orifice installed in the system is:

The flow rate equals the area of the orifice in square inches multiplied by the gallon-per-minute flow rate times the unit of measure adjustment factor of 0.3208.

As with all formulas, the units of measurement must be consistent. Let us take an example of a system that is flowing at 1 gallon per minute. This is equivalent to 231 cubic inches per minute. Let us also say that we are going to force the fluid through an orifice of θ 0.25 inches. The Greek letter θ (theta) refers to the diameter of the orifice. To find the area of the orifice, we square the radius and multiply it by pi. Therefore, the area of the 0.25 orifice is:

$$0.250/2 = 0.125$$
$$0.125 \times 0.125 = 0.016$$
$$0.016 \times 3.14 = 0.05 \text{ square inches}$$

One gallon per minute though an orifice of 0.0079 square inches would yield a velocity of 6.4 feet per second through the orifice. This results in an increase in velocity that causes a drop in pressure according to Bernoulli's law.

The pressure drop in this example can be estimated using the following formula:

$$\text{Pressure Drop } (\Delta P) = [GPM/(23.5 \times A)]^2$$

The constant of 23.5 in the preceding formula is a number that was derived experimentally by measuring pressure drops across typical orifices. For that reason and because of such variables as temperature, oil viscosity, oil specific gravity, the sharpness of the edge on the orifice, and also the characteristics of the plumbing prior to the orifice and after the orifice, this formula should only be used to provide a rough estimate of the amount of pressure drop.

In the previous example, the gallon per minute flow was 1 and the area of the orifice was 0.05 in^2.

$$[1/(23.5 \times 0.05)]^2 = [1/(1.175)]^2 = [0.851]^2 = 2.67$$

STRAIGHT THREAD FITTING SIZES

Most of the fittings used in hydraulic systems use a straight thread fastening system to connect lines to components. Identification codes found on many pre-built hoses are used to identify not only the hose but also the fitting. Fitting sizes demand the use of a specific size O-ring **(see Table 7-5)**.

THREAD FORMS OF FLUID CONNECTORS

National Pipe Thread Fuel (NPTF)

- Thread conforms to ANSI B1.20.3.
- Physically interchangeable with NPT but has modified threads for better pressure tight sealing.

TABLE 7-5: STRAIGHT THREAD FITTING SIZES					
Fitting Dash No.	Tubing O.D.	Thread Size	ARP 568 Uniform Dash No.	O-Ring I.D.	O-Ring Thickness
−2	1/8″	5/16–24	−902	0.239	0.064
−3	3/16″	3/8–24	−903	0.301	0.064
−4	1/4″	7/16–20	−904	0.351	0.072
−5	5/16″	1/2–20	−905	0.414	0.072
−6	3/8″	9/16–18	−906	0.468	0.078
−8	1/2″	3/4–16	−908	0.644	0.087
−10	5/8″	7/8–14	−910	0.755	0.097
−12	3/4″	1 1/16–12	−912	0.924	0.116
−14	7/8″	1 3/16–12	−914	1.048	0.116
−16	1″	1 5/16–12	−916	1.171	0.116
−20	1 1/4″	1 5/8–12	−920	1.475	0.118
−24	1 1/2″	1 7/8–12	−924	1.720	0.118
−32	2″	2 1/2–12	−932	2.337	0.118

- Tapered thread profile seals by metal-to-metal interference fit; usually requires a sealing compound for pressure tight connections.
- Pitch and diameter are measured in inches.
- Taper angle is 0.75" per foot or 1 degree 47 minutes.
- Thread angle is 60 degrees.

Straight Thread O-Ring (SAE)

- Thread conforms to ISO 263 and ANSI B1.1 Unified.
- Port conforms to ISO 11926 and SAE J1926.
- Commonly called straight thread a-ring fittings.
- Pitch and diameter are measured in inches, 1\116-12 UN-2B.
- Threads are parallel and require a-ring for a pressure tight connection.
- Thread angle is 60 degrees.

British Standard Pipe Tapered (BSPT)

- Thread conforms to ISO 7.
- Pitch and diameter are measured in inches—for example, G3/8-19.
- Tapered thread profile seals by metal-to-metal interference fit; usually requires a sealing compound for pressure tight connections.
- Taper angle is 1 degree 47 minutes, the same as NPT(F).
- Thread angle is 55 degrees.
- Not interchangeable with NPT(F).

British Standard Pipe Parallel (BSPP)

- Thread conforms to ISO 228-1.
- Port conforms to ISO 11 79.

- Pitch and diameter measured in inches—for example, G1I4-19.
- Parallel threads require O-ring, crush washer, gasket, or metal-to-metal seal between connections for pressure tight connections.
- Thread angle is 55 degrees.
- Not interchangeable with SAE or NPT(F).

Metric Straight Thread O-Ring

- Thread conforms to ISO 261.
- Port conforms to ISO 6149 and SAE J2244.
- Pitch and diameter measured in millimeters—for example, M22 × 1.5.
- Parallel threads require O-rings for pressure tight connections.
- Thread angle is 60 degrees.
- Easily identified by the raised ridge on the female port counterbore.
- Not interchangeable with SAE or BSPP.

ISO STANDARDIZATION

The **International Standards Organization (ISO)** is attempting to bring together standards from around the world, with the goal of developing a limited number of standards for fluid power equipment around the world. The end result of this, once successful, is that the service technician will need to stock or have access to a smaller variety of components in order to successfully maintain and repair mobile fluid power equipment. **Table 7-6** illustrates how efforts are underway to combine **American National Standards Institute (ANSI)** standards with ISO standards to simplify hydraulic hoses for the global marketplace.

TABLE 7-6: ANSI TO ISO STANDARDIZATION					
Application	Port	24° Cone Bite Type	37° Flare	Metric ORFS	24° Cone Weld Nipple
For All Designs	Metric ISO 6149 (SAE J2244)	ISO 8434-1	ISO 8434-2	ISO 8434-3	ISO 8434-4
Not for New Designs in Hydraulic Fluid Power Systems	BSPP ISO 1179 (DIN 3852-2)	ISO 8434-1	ISO 8434-2		ISO 8434-4
	Metric ISO 9974 (DIN 3852-1)	ISO 8434-1			ISO 8434-4
	UN/UNF ISO 11926 (SAE J1926)		ISO 8434-2		

Summary

When first getting acquainted with categorizing and ordering hydraulic hoses, the novice may become overwhelmed with the terminology and variations. This is where seeking the assistance of a more experienced technician or parts professional can provide both help and an education. Often even the experienced technician will need to remove the hose from the equipment and take or send to the supplier to ensure that the correct hose is provided for replacement. Since hoses are often identified by inside diameter, length, strength, and type of fitting, it is essential that the service technician be familiar with these measurements and classifications to ensure the correct hose or tubing is ordered and installed.

Review Questions

1. Technician A says that any fitting is acceptable as long as the fitting is tightened using the proper torque. Technician B says that it is virtually impossible to overtorque the fittings on hydraulic hoses and pipes. Who is correct?

    A. Technician A

    B. Technician B

    C. Technician A and Technician B

    D. Neither Technician A nor Technician B

2. A technician is tightening a fitting, the size of which is −06. Unable to find the proper specification in the data supplied by the fluid power system's manufacturer, the technician goes to the Internet to look up generic torque data for the size fitting. Concerned about the extra time being taken, the technician's manager tells the technician to simply tighten the fitting until it is finger type and then torque it at 90 degrees. At the same time, the technician discovers that the generic torque spec for this fitting is 18 to 20 pounds-feet of torque. Who is correct?

    A. The technician is correct because he or she verified the proper torque specification.

    B. The manager is correct because from experience he or she knows that the rule of thumb for torquing this size and type of fitting is one half **flats** of the fitting's hex surfaces.

    C. Both are correct because rules of thumb are as good as specifications.

    D. Neither is correct because the specification is actually far different from what either of them knows or has found.

3. Which of the following provides the best seal in a high-pressure fluid power system?

    A. The tapered metal-to-metal seal

    B. The straight metal-to-metal seal

    C. The 30-degree cone seal

    D. The O-ring face seal

4. The split flange O-ring seal is typically used to connect hoses to pumps and hydraulic motors. One of the advantages of this type of seal is:

    A. There is no spiral path for leaks.

    B. Any type of sealing compound can be used on the O-rings.

    C. Burrs, nicks, dirt, and other foreign materials will not affect the sealing ability of this type of fitting.

    D. Torque is not critical.

5. On a fluid power diagram, the Greek letter theta (θ) represents:
   A. The length of an orifice.   C. The diameter of an orifice.
   B. The thread pitch on a fitting. D. The area of an orifice.

6. Which of the following is the fitting shown in the figure above?
   A. A 37-degree JIC   C. An SAE 45-degree cone seal
   B. An O-ring face seal   D. A split flange O-ring seal

7. Which of the following is the fitting shown in the figure above?
   A. A 37-degree JIC   C. An SAE 45-degree cone seal
   B. An O-ring face seal   D. A split flange O-ring seal

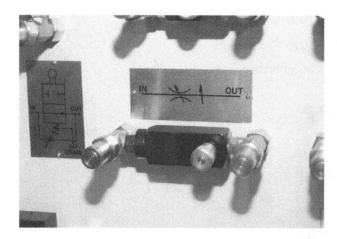

8. Which of the following is the fitting shown in the figure above?

   A. A 37-degree JIC

   B. A quick disconnect

   C. An SAE 45-degree cone seal

   D. A split flange O-ring seal

9. The adjustable orifice shown in the figure above is adjustable to a thousandth of an inch. The current setting on this device is:

   A. Approximately 0.535.

   B. Approximately 0.350.

   C. Approximately 0.355.

   D. Approximately 0.00535.

10. Which of the following formulas is correct for calculating pressure drop?

    A. $(\Delta P) = $ Rate $\times$ Distance / Time

    B. $(\Delta P) = [GPM / (23.5 \times A)]^2$

    C. $(\Delta P) = GPM \times A^2$

    D. It cannot be calculated.

8 System Design

Learning Objectives

Upon completion and review of this chapter, the student should be able to:

- Analyze the operation of hydraulic fluid power systems based on hydraulic diagrams provided by the manufacturer.
- Use hydraulic diagrams to diagnose and troubleshoot fluid power systems.

Cautions for This Chapter

- A malfunctioning fluid power system can experience high temperatures at various points and in various components throughout the system. The temperature of these components should not be checked by touch, but rather by an infrared heat gun.

- A malfunctioning fluid power system may have weak hoses and lines in the system. The technician working on that in a functioning system should always wear protective eyewear, clothing, and gloves to reduce the chances of injury.

- A malfunctioning fluid power system can suffer from catastrophic leaks at a time. While the system is operating, and especially while testing the system, the technician should protect himself or herself from direct exposure to lines and components that suddenly rupture.

- Hydraulic fluid under pressure can be injected through the skin. The technician working on a malfunctioning system should never place his or her hands or other body parts on or near any components while testing the system.

- Before affecting any repairs, the service technician should follow proper lock out and tag out protocols as outlined by the safety officers within his or her organization.

- Spilled oil can be very slick and hazardous to walk on or handle equipment on.

Key Terms

directional control valve

pressure relief valve

solenoid-operated valve

tandem

INTRODUCTION

All fluid power systems are designed to push, pull, lift, lower, and rotate. When designing a system, the engineer or technician needs to first determine what task the system needs to perform. More complex systems may need to do all five of these tasks. For some technicians this can create system diagrams that are overwhelming. What the technician must remember is that the defect will generally exist in only one component. This is not to say that there cannot be collateral damage, but the beginning of the failure can usually be found in a single component. An important skill for the technician is to understand fluid power systems on a design basis. Each component has a specific task and relates to every other component in a specific manner. As the technician gathers information to begin a diagnosis, he or she should be thinking in terms of what part of the system is not working, what that part of the system is supposed to do, and how that part of the system relates to the machine as a whole.

In general, two terms are used when discussing a group of hydraulic components working together in unison. A hydraulic system is a group of interconnected hydraulic components mounted on a vehicle or a piece of equipment. For example, a cement mixer will have a relatively complex assemblage of components designed to rotate the hopper, perhaps tell the hopper, perhaps operate components on the chute, and perhaps pump the cement from the hopper. A hydraulic circuit is that portion of the hydraulic system that is designed to perform a specific task. In most cases, the hydraulic system will have a single pump in a single reservoir that supplies fluid for several circuits; each one of the circuits is designed to perform a specific task. Therefore, if a technician studies the circuit that is not performing its assigned task properly, than troubleshooting and repair will not be overwhelming.

Starting with the most fundamental aspect of how fluid power systems works, three categories of circuits exist:

- Open center circuits
- Closed center circuits
- Closed loop circuits

OPEN CENTER CIRCUITS

Open center hydraulic circuits are so named because of the type of **directional control valve** that is used. An open center control valve provides the passageway so that when the control valve is in the neutral position, oil passing through the valve from the pump is directed back to the reservoir. As was covered in an

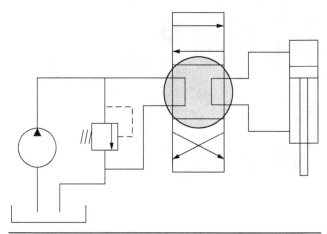

Figure 8-1 Systems are often referred to as either open center or closed center. This refers to what happens to the fluid when a directional control valve is in a neutral or center position. In an open center system, the fluid passes through the directional control valve and returns back to the reservoir on one side and the actuator on the other side. In this system, there is no pressure on the supply side when idle, and fluid is able to flow easily in one direction through the actuator. In the case of a piston, the piston is free to either extend or retract depending on the force applied to it.

earlier chapter, there are both open center and closed center directional control valves. Additionally, the center position on a three-position directional control valve can be open center on the inbound and closed center on the outbound side, or vice versa. When talking about a circuit being of an open center design, this refers to the inbound side only. A circuit that is closed on the inbound side and open on the outbound side would be referred to as a closed center circuit **(Figure 8-1)**.

The advantage in a simple circuit is that the pump is drawing a little power from the engine or motor when the directional control valve for the circuit is in the neutral position. Therefore, the pump can remain engaged, which simplifies the control mechanism for driving the pump. However, hydraulic systems utilizing multiple circuits in an open center circuit also have a disadvantage. When the circuit utilizing the open center directional control valve is in the neutral position, fluid being supplied by the pump has almost no resistance in returning back to the reservoir. This means that parallel circuits will have little to no pressure and flow available to them.

Since open center circuits are usually found in relatively simple systems, they typically will use relatively simple fixed displacement pumps. Remember that a fixed displacement pump moves the same volume of fluid in a given speed regardless of the amount of pressure that is moving against the fluid. Therefore, when a restriction occurs or when an actuator reaches a maximum load condition, there must be a method of relieving the excess pressure. This is done with the **pressure relief valve**. The excess pressure then returns back to the reservoir by way of the primary filter.

Open center circuits are the least complex of all hydraulic systems found on mobile equipment. They are the simplest to troubleshoot and tend to use fewer and relatively inexpensive components.

CLOSED CENTER CIRCUITS

In a **closed center** system **(Figure 8-2)**, the directional control valve does not provide a pathway back to the reservoir when the control valve is in the neutral position. Like open center circuits, closed center circuits can be very simple; however, they tend to be found in more complex mobile hydraulic systems. Since the oil flow through the directional control valve is blocked when the valve is in the neutral position, there is no place for the oil to flow. Therefore, there must be another part of the system, and another circuit, through which the oil will travel and ultimately find its way back to the reservoir. In some cases, the equipment will

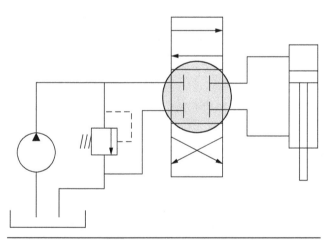

Figure 8-2 The opposite of an open center system is a closed center system. When the directional control valve in a closed center system is placed in the idle, neutral, or center position, then all fluid flow stops. Fluid coming from the pump is stopped or "dead headed" at the directional control valve. Because the fluid flow from the pump is stopped, pressure between the pump in the directional control valve would go to the maximum capacity of the pump. To prevent this, there will always be a pressure relief valve parallel to the line between the pump and the directional control valve. On the actuator side of the directional control valve, fluid flow is also stopped. This would cause the actuator, the motor, or the cylinder to be parked in its last position and unable to move.

be designed so that when one directional control valve is in the neutral position, the other directional control valve will be in an open flow position. It may also mean that when one directional control valve is in the neutral position and the other directional control valve is also in the neutral position, oil returns to the reservoir through a pressure relief valve. This latter scenario means that the pump is always pumping at a high pressure and this tends to be wasteful of the energy being supplied to rotate the pump by the power source, the diesel engine, the gasoline engine, or the electric motor.

The advantage of a closed center circuit is that it does allow flow and pressure to be available to other circuits in the system when that circuit's control valve is in the neutral position. This makes closed center directional control valves ideal for more complex systems. Another distinct advantage of the closed center directional control valve is that it does allow for an accumulator to be charged while the valve is in the neutral position.

CLOSED LOOP SYSTEMS

In both an open center circuit and a closed center circuit, the oil being used by the hydraulic circuit returns to the reservoir, where it rests for a while before being picked up by the pump again. While resting, it has an opportunity to cool, and has an opportunity for foreign objects in contamination to settle to the bottom of the reservoir. These systems are both referred to as open loop systems. Closed loop systems are not found in a cylinder actuator system, only in a hydraulic motor system. In a closed loop system, assuming we live in a perfect universe, the pump would move fluid to the motor, the fluid would pass through the motor, and then be returned to the pump to be immediately directed back to the motor. If the motor operated at 100 percent volumetric efficiency, this could work. It also means that there would be no oil available to lubricate the pump. Therefore, a more believable volumetric efficiency would be about 90 percent. This means that 10 percent of the volume of fluid moved by the pump through the motor would be used to lubricate components and then would be returned back to a reservoir through a line called a drain. With this 10 percent loss, eventually the closed loop circuit would run out of oil. Therefore, a secondary pump known as the charge pump picks up fluid from the reservoir and adds it into the closed loop **(Figure 8-3)**. This ensures that the pump does not cavitate and that the motor gets adequate flow and pressure to rotate at the proper rate and with the proper amount of power. These closed loop systems are the core of hydrostatic transmissions and hydrostatic drive circuits.

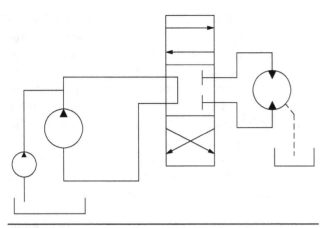

Figure 8-3 There is also a category of systems known as closed loop systems. In these systems, fluid is pumped directly to a motor and then returned directly back to the pump. What flows from a pump passes through the motor, does not return to a reservoir, and is simply recycled through the pump back to the motor once again. Because of this, the fluid required for lubrication will always be a small bleed off from the motor to the reservoir. Thus, there is often a secondary pump that will inject a small amount of hydraulic fluid from the reservoir back into the system to provide the extra fluid required for lubrication.

Accumulator Sizing

An accumulator **(Figure 8-4)** has three primary purposes in a system. One is to smooth out pulsations to allow for a more even flow, and therefore smoother operation of the actuators. Another use of an accumulator is to act as a shock absorber. Accumulators are also used to store energy.

Accumulators are basically metal cylinders with the dividing diaphragm in them. One side of the diaphragm is open and basically contains the air at atmospheric pressure prior to being installed in the system. The other side of the diaphragm has a gas charge of approximately half the maximum system or circuit operating pressure. Since most systems in mobile equipment applications operate at pressures no higher than 3,000 psi, the nitrogen charge in most accumulators used in mobile applications will be about 1,500 psi. The accumulators are typically rated by the manufacturer and gas volume when all the gas has been expelled.

Accumulators are selected based on the job they have to do in the circuit. For instance, if the job is simply to smooth out relatively minor pulsations in pressure or flow, then the accumulator will probably be relatively small **(Figure 8-5)**. If the accumulator is used to ensure the rapid movement of a cylinder, it will be large. For instance, there is an American-built piece of military equipment that has a hydraulically operated door

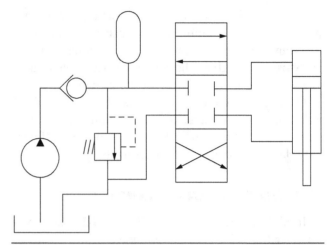

Figure 8-4 Accumulators are used in chiefly two ways. The first way is to act as a buffer to smooth out pressure pulsations coming from the pump or other components in a fluid power system. The second way in which an accumulator can be used is to store a volume of fluid under pressure. In this example, the accumulator is filled and pressurized while the directional control valve is in a neutral or closed center position. When the directional control valve is moved to either the straight flow position or the cross flow position, the accumulator discharges toward the cylinder. The end result is a rush of fluid flow that far exceeds the capacity of the pump. Therefore, the piston is moved more rapidly than the pump could do on its own.

between the crew compartment and the ammunition storage area. It is important to close this door quite quickly. When the door is open, an accumulator with a volume approximately twice that of the fully extended cylinder operating the door is fully pressurized. When the door is to be shut, a valve is opened allowing the contents of the accumulator to flow into the door's cylinder at a very high volume per minute, which causes the door to close in a fraction of a second.

Several formulas for accumulator sizing are available and are used by engineers when designing a system, circuit, or hydraulic system. The general formula for most accumulators is:

$$D = (e \times P_1 \times V_1)/P_2 - (e \times P_1 \times V_1)/P_3$$

D = **volume of fluid discharge (in³)**

P_1 = **pre-charge pressure (psi)**

P_2 = **system pressure after volume D has been discharged**

P_3 = **maximum system pressure at full accumulator pressure**

V_1 = **rated accumulator gas volume (in³)**

e = **system efficiency, typically 0.95**

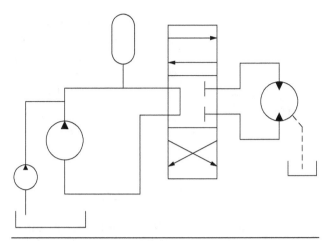

Figure 8-5 In this example, the accumulator is used to smooth out pressure pulse stations originating at the pump. With the open center directional control valve in the neutral position, fluid is flowing from the pump past the accumulator in returning back to the pump. When the directional control valve is moved to either the straight flow position or the cross flow position, the fluid flows to the motor allowing the motor to turn. If there is work to do as the motor begins to rotate, the pressure on the inbound side of the motor will increase. As the pressure increases, the pump will charge the accumulator. Should there be a drop in pressure during rotation of the motor, the fluid stored in the accumulator will move back into the main flow of fluid to the motor and bring the pressure in flow rate back to the average level. The end result is smoother operation of the hydraulic motor as the load on the hydraulic motor increases and decreases.

In word format, this would be:

D = (system efficiency times pre-charge pressure times rated accumulator gas volume in cubic inches) divided by system pressure after the volume of fluid discharge in cubic inches minus (system efficiency times pre-charge pressure times rated accumulator gas volume in cubic inches) divided by maximum system pressure at full accumulator pressure.

Let us say that we want to extend the cylinder, which requires 1 gallon of hydraulic oil to extend it through its full range. As we know from Chapter 1, a U.S. gallon = 231 cubic inches. Therefore, the discharge volume required (D) is 231 cubic inches of hydraulic oil.

The first step in calculating the size accumulator that is required is to calculate the volume of the uncharged, or pre-charged, accumulator. If the system efficiency is 95 percent, which is typical, then 0.95 should be multiplied by the pre-charge pressure and the rated accumulator gas volume in cubic inches. If the pre-charge

gas pressure is 1,500 psi and the pre-charge volume of the accumulator is 300 cubic inches, then our first step yields 427,500 divided by the system pressure after the accumulator has been discharged. For discussion purposes, we will say that this pressure needs to be 500 psi. The final calculation of the first step is 855.

The next step is to calculate the volume of the accumulator when the accumulator is charged the full accumulator pressure. Again, the system efficiency is multiplied times the pre-charge pressure and the rated accumulator gas volume. This again would yield 427,500. This time, however, we will multiply by the pressure at full accumulator system pressure, which we will say for this example is 2,500 psi. Thus, the final calculation for the second step is 171.

The final step is to subtract the second calculation, which ended up being 171, from the first calculation, which was 855. This yields a result of 684. We only required 231 cubic inches (remember that hydraulic oil is not compressible), therefore this accumulator is actually almost three times larger than what is required.

One might suggest that natural large accumulators would be a good thing, and in some cases that might be correct, but in most cases the extra expense of the larger accumulator is not worth any benefit that might be gained. The proper size accumulator would be considerably smaller. However, as the designer, when the system reaches the optimum size for the accumulator, it is advised that they add approximately a 5 percent extra capacity.

REQUIRED FLOW FOR OPERATING A HYDRAULIC CYLINDER

A previous chapter discussed how to calculate flow rates. When sizing a pump it is necessary to know the size of the cylinder and how long it takes for the cylinder to go from the fully retracted position to the fully extended position **(Figure 8-6)**. This formula in gallons per minute is:

(area times stroke times 60) divided by (time times 231)

Or

(area × stroke × 60) ÷ (time × 231)

Therefore, if a 4-inch cylinder has a stroke of 18 inches and needs to travel that distance in 5 seconds, the calculation would be:

(12.56 × 18 × 60) = 6,028.8

(5 × 231) = 105

6,028 ÷ 105 = 5.790 gallons per minute

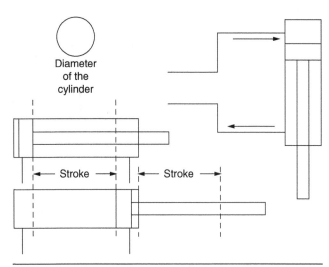

Figure 8-6 The volume of the cylinder is found using a mathematical formula: half the diameter of the cylinder times pi times the length of the stroke of the cylinder. This yields a result in cubic inches. Volume is sometimes given in cubic inches, sometimes given in liters, and sometimes given in gallons. If the specification is given in gallons, it is important to look at the manufacturer of the equipment. If the equipment has been designed or built in Great Britain, Canada, or Australia, the gallons unit of measure is likely to be Imperial gallons. The Imperial gallon is larger than the U.S. gallon, and therefore can cause calculations to be incorrect if the assumption is that the specification is in U.S. gallons.

HORSEPOWER TO DRIVE A PUMP

The amount of horsepower required to drive a hydraulic pump will vary with pump efficiency, with the flow rate, and with the pressure the system is attempting to draw from the pump **(Figure 8-7)**. A crude rule of thumb suggests that one horsepower is required for each gallon per minute at 1,500 psi.

Since most mobile fluid power systems use a diesel engine as a power source to drive the pump, it is necessary to remember that operating the system takes power from the engine. The horsepower is equal to the pressure times the gallons per minute divided by 1,714 times the efficiency factor, which is usually about 85 percent (0.85).

$$\textbf{HP} = \textbf{PSI} \times \textbf{GPM} \div (\textbf{1,714} \times \textbf{0.85})$$

This means at 1,000 psi, the flow rate of 5 gallons per minute in the fluid power system will be a load on the engine equivalent to 3.43 HP. For most diesel engines found in trucks, this is a negligible amount even at idle. However, increase the gallons per minute to 100 and the pressure to 5,000 psi and the load of the fluid power system rises to 343 HP, which

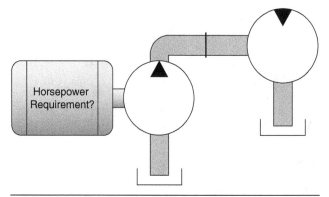

Figure 8-7 Any fluid power system requires an outside power source to drive the pump. In some cases, this pump is an electric motor; in others, it is a gasoline engine. In many cases, it is a diesel engine. The amount of work the fluid power system must accomplish means that the same amount of power must be provided, or be taken away from, this power source. Many people have experienced the engine of their car stalling while attempting to parallel park. Sometimes, especially if it occurs when the steering wheel is being turned, this is caused by the power steering pump, which is nothing more than a hydraulic pump, drawing more power from the engine in order to operate the pump (which in turn supplies power to the hydraulic actuator in the steering) then the engine is able to supply. For operating in a fluid power system, a minimal amount of horsepower must be drawn from the power source. This must at least be equal to the amount of power required by the actuator, cylinder, or motor in the fluid power system.

would be a significant load on almost any truck engine. Critically important is to realize that this load of 343 HP generates significant heat. Thus, within the engine itself the cooling package must be large enough and able to dissipate enough heat to generate this power without overheating and damaging the engine. Increase the heat load on the cooling package of the engine with an ambient air temperature like one might find in Phoenix, Arizona, in the middle of summer and the cooling package may not be able to keep up.

Table 8-1 is a quick reference chart that will work well for most truck-mounted systems. Many of the two samples above the pressure were increased by a factor of 5, and the oil flow rate was increased by a factor of 20. This gave an overall system first power increase that was 100 times higher (5 × 20). Not surprisingly, the horsepower load was an increase of 100 times. This chart can therefore be used as a basis for much larger flow rates, much larger pressures, and much larger engines.

	TABLE 8-1: HORSEPOWER REQUIRED TO DRIVE A PUMP									
GPM	500 PSI	750 PSI	1,000 PSI	1,250 PSI	1,500 PSI	1,750 PSI	2,000 PSI	2,500 PSI	3,000 PSI	5,000 PSI
1/2	0.172	0.257	0.343	0.429	0.515	0.6	0.686	0.858	1.03	1.72
1	0.343	0.515	0.686	0.858	1.03	1.2	1.37	1.72	2.06	3.43
1 1/2	0.515	0.772	1.03	1.29	1.54	1.8	2.06	2.57	3.09	5.15
2	0.686	1.03	1.37	1.72	2.06	2.4	2.75	3.43	4.12	6.86
2 1/2	0.858	1.29	1.72	2.14	2.57	3	3.43	4.29	5.15	8.58
3	1.03	1.54	2.06	2.57	3.09	3.6	4.12	5.15	6.18	10.3
3 1/2	1.2	1.8	2.4	3	3.6	4.2	4.8	6	7.21	12
4	1.37	2.06	2.75	3.43	4.12	4.8	5.49	6.86	8.24	13.7
5	1.72	2.57	3.43	4.29	5.15	6	6.86	8.58	10.3	17.2
6	2.06	3.09	4.12	5.15	6.18	7.21	8.24	10.3	12.4	20.6
7	2.4	3.6	4.8	6	7.21	8.41	9.61	12	14.4	24
8	2.75	4.12	5.49	6.86	8.24	9.61	11	13.7	16.5	27.5
9	3.09	4.63	6.18	7.72	9.27	10.8	12.4	15.4	18.5	30.9
10	3.43	5.15	6.86	8.58	10.3	12	13.7	17.2	20.6	34.3
12	4.12	6.18	8.24	10.3	12.4	14.4	16.5	20.6	24.7	41.2
15	5.15	7.72	10.3	12.9	15.4	18	20.6	25.7	30.9	51.5
20	6.86	10.3	13.7	17.2	20.6	24	27.5	34.3	41.2	68.6
25	8.58	12.9	17.2	21.4	25.7	30	34.3	42.9	51.5	85.8
30	10.3	15.4	20.6	25.7	30.9	36	41.2	51.5	61.8	103
35	12	18	24	30	36	42	48	60	72.1	120
40	13.7	20.6	27.5	34.3	41.2	48	54.9	68.6	82.4	137
45	15.4	23.2	30.9	38.6	46.3	54.1	61.8	77.2	92.7	154
50	17.2	25.7	34.3	42.9	51.5	60	68.6	85.8	103	172
55	18.9	28.3	37.8	47.2	56.6	66.1	75.5	94.4	113	189
60	20.6	30.9	41.2	51.5	61.8	72.1	82.4	103	124	206
65	22.3	33.5	44.6	55.8	66.9	78.1	89.2	112	134	223
70	24	36	48	60	72.1	84.1	96.1	120	144	240
75	25.7	38.6	51.5	64.3	77.2	90.1	103	129	154	257
80	27.5	41.2	54.9	68.6	82.4	96.1	110	137	165	275
85	29.2	43.8	58.3	72.9	87.5	102	117	146	175	292
90	30.9	46.3	61.8	77.2	92.7	108	124	154	185	309
95	32.6	48.9	65.2	81.5	97.8	114	130	163	196	326
100	34.3	51.5	68.6	85.8	103	120	137	172	206	343

PUMP AND MOTOR TORQUE

While horsepower is an important factor in the operation of the cylinder and the sustained operation of a motor moving a load, torque is an important part of the motor's ability to get a load moving by means of its rotation (**Figure 8-8**). Calculating pump and motor torque is done by multiplying the diameter of the pump or motor times the pressure and dividing that by 24 times pi.

$$T = (D \times PSI) \div 24\pi$$

The torque required to get a load moving is a function of the weight and the radius of rotation. For instance, if the hydraulic motor is rotating a winch that has a radius twice that of the motor, the actual load on both the motor and the pump is twice that of the load on the drum. Let us say the diameter of the drum is 18 inches and the diameter of the motor is 9 inches. If it takes 100 pound-feet of torque to get the load moving at the circumference of the drum, it will take 200 pound-feet of torque at the pump to begin moving that load. See **Table 8-2** for an easy estimate of torque based on flow and pressure.

TABLE 8-2: PUMP AND MOTOR TORQUE (IN POUND/FEET)

GPM @ R1200 RPM	DISPL. Per C.I.R.*	250 PSI	500 PSI	750 PSI	1,000 PSI	1,250 PSI	1,500 PSI	2,000 PSI	2,500 PSI	3,000 PSI
3	0.577	1.91	3.82	5.74	7.65	9.57	11.5	15.3	19.1	23
5	0.962	3.19	6.38	9.56	12.7	15.9	19.1	25.5	31.9	38.2
8	1.54	5.22	10.4	15.7	20.9	26.1	31.3	41.8	52.2	62.7
10	1.92	6.37	12.7	19.1	25.5	31.9	38.2	51	63.7	76.5
12	2.31	7.64	15.3	22.9	30.6	38.2	45.8	61.1	76.4	91.6
18	3.46	11.4	22.9	34.4	45.9	57.4	68.8	91.7	115	138
25	4.81	15.9	31.9	47.9	63.9	79.7	95.7	127	159	191
40	7.7	25.5	51.1	76.5	102	127	153	204	255	306
50	9.62	31.9	63.8	95.6	127	159	191	255	319	382
75	14.43	47.8	95.6	144	191	239	287	383	478	574
85	16.43	54.2	108	163	217	271	325	434	542	651
100	19.2	63.7	127	191	255	319	382	510	637	765

*C.I.R. = cubic inches per revolution

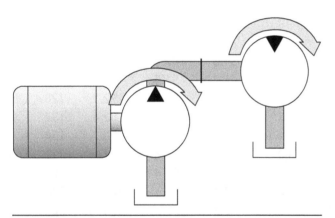

Figure 8-8 Torque is important for a hydraulic motor to begin lifting a load or to change the speed at which a load is being lifted. The amount of torque the motor requires must be matched by the amount of torque generated by the hydraulic pump. Therefore, if the amount of torque required by the motor is known, then the amount of torque required of the pump can easily be calculated.

Mechanical Transmission Efficiency

A hydraulic motor that is connected mechanically through a power transmission device must be able to supply enough additional power to overcome the loss of the transmission device. In a complex piece of machinery, there may be several transmission links. For instance, the motor might drive the belt, the belt then drives a helical gear reducer, which in turn drives a cam reaction drive. The transmission efficiency loss would have to be calculated for each stage in transmitting the power from the hydraulic motor to the load. **Table 8-3** shows approximate transmission losses for each type of machine.

TABLE 8-3: POWER TRANSMISSION EFFICIENCY BY TYPE

Power Transmission Type	Typical Efficiency
V-belt drives	95%
Gear cog belt drives	98%
Poly-V or ribbed belt drives	97%
Flat belt drives, leather, or rubber	98%
Flat belt drives, nylon core	90% to 99%
Variable speed, spring-loaded, wide range:	
V-belt drives	80% to 90%
Compound drive	75% to 90%
Cam-reaction drive	95%
Helical gear reducer:	
Single stage	98%
Two-stage	96%
Worm gear reducer:	
10:1 ratio	86%
25:1 ratio	82%
60:1 ratio	66%
Roller chain	98%
Leadscrew, 60° helix angle	65% to 85%
Flexible coupling, shear type	99% plus

Figures 8-9 through **8-28** show a system being designed and built from the simplest to a relatively complex design. Study these drawings along with the captions to gain a better understanding of how components work together in a fluid power system to perform a specific task or job.

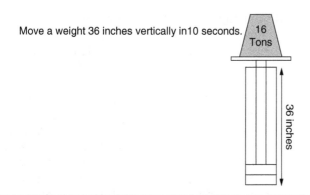

Move a weight 36 inches vertically in10 seconds.

Figure 8-9 Any mass can be moved by virtual pressure above atmospheric pressure. The lower the pressure, the larger the area of the piston must be to move a given weight. Sixteen tons can be moved by as little as 1 psi (or even less). To do so, the surface area would have to be a little over 32,000 in². If we want to move this mass 36 inches vertically, the volume required to do so would be 1,152,000 in³. That equates to nearly 5,000 U.S. gallons. If time is a factor in this, such as a need to have the weight moved from the full retract position to the full extent position of the piston in 10 seconds, then the flow rate would need to be 30,000 gallons per minute. This is an absurdly high flow rate, however. Determining the proper cylinder size is therefore a function of how fast the mass needs to be moved versus the pressure at which the designer wants to operate.

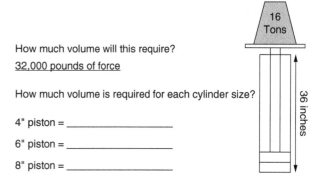

How much volume will this require?
32,000 pounds of force

How much volume is required for each cylinder size?

4" piston = _____

6" piston = _____

8" piston = _____

Figure 8-11 In this example, we are being asked to move the 32,000-pound mass 36 inches. How much volume required to move this mass 36 inches can be calculated by multiplying the surface area (in square inches) by the distance the mass needs to be moved. Even though the larger piston requires a lower operating pressure, it will require a larger volume to move the mass through the designated distance. The higher pressures require less flow, but also require components made out of a more robust material. It is a simple fact that the more robust material, the more expensive the components will be. Designing a hydraulic system is always a function of weighing the cost of the system against the size of the components, as well as the practicality of using high pressures versus low pressures.

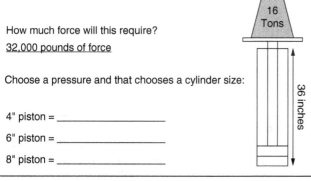

How much force will this require?
32,000 pounds of force

Choose a pressure and that chooses a cylinder size:

4" piston = _____

6" piston = _____

8" piston = _____

Figure 8-10 Although many limiting factors exist in the design of a fluid power system, the key factors are the desired operating pressure and the strength—and therefore the expense—of the metallurgy involved in the components. The stronger the components, the higher the operating pressures can be, and therefore the smaller the components and the smaller the amount of fluid required. In the example shown here, determine how much pressure is required to move the piston exerting a force on the 32,000-pound weight vertically. Since the force requires 32,000 pounds, simply divide that factor by the surface area of each of the three pistons listed. This will tell you the pressure required for each piston size. Note that the smaller the piston, the higher the required pressure.

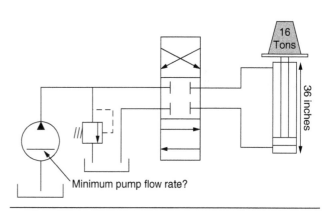

Minimum pump flow rate?

Figure 8-12 Choose one of the piston sizes from the previous illustration. What will the flow rate need to be for the piston size chosen in order to accomplish the 36-inch move in 10 seconds? This will be calculated by multiplying the volume required to move the piston the required length times the fraction of a minute that 10 seconds represents. In this case, 10 seconds is 1/6 of an minute, so the volume will be multiplied times six in order to determine the flow rate. The smaller the piston, the lower the required flow rate; the larger the piston, the higher the required flow rate.

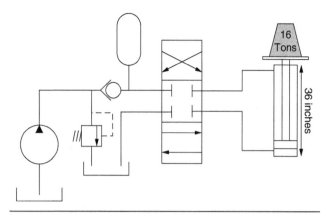

Figure 8-13 When an engineer designs a fluid power system, it may be decided to use accumulators to reduce the length of time it takes for the piston to move through the required range. Engineers will use certain formulas to determine the proper size of the accumulator to achieve the desired result. When making a repair to a system, it is critically important to replace the accumulator with the size specified by the designer of the system. It is equally important that the pre-charge in the accumulators, that is to say the pressure of the nitrogen charge behind the accumulator's diaphragm, be at the designer's recommendation. In the design shown here, the pump will flow and the pressure relief valve will remain closed until the accumulator is charged to the pressure set by the pressure relief valve when the directional control valve is in the center position. With the accumulator charged, the check valve prevents flow from the accumulator back to the pump, and the directional control valve prevents flow of the fluid toward the cylinder. When the directional control valve is opened either to the straight flow or cross flow position, then the fluid can flow toward the cylinder, thereby draining the accumulator. The speed of discharge of the accumulator will be limited only by the restriction to flow created by the directional control valve and the hosing to and from the cylinder.

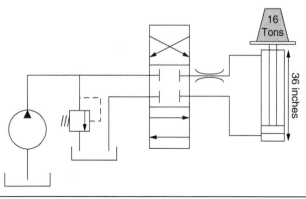

Figure 8-14 In this illustration, an orifice has been added to the line leading from the retract side of

the cylinder to the directional control valve. This is not the limit of the speed at which the piston will retract but rather the speed at which the piston will extend. In a straight flow position of the virtual control valve, the fluid moves from the pump through the orifice and into the retract side of the cylinder. Although this does begin to move the piston downward, in reality the 32,000-pound mass is going to have a much greater impact on the movement of the piston than will the fluid from the pump. The piston will in fact move very rapidly to the full retract position because of the mass. In the cross flow position, fluid from the pump passes to the extend side of the piston. Although there is no flow control orifice on the extend side, fluid cannot enter the extend side until fluid leaves the retract side. Therefore, limiting the flow rate of the fluid on the retract side will also limit the rate at which fluid can pass into the extend side.

Flow through an orifice to slow descent

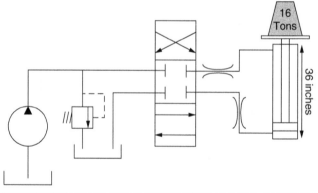

Figure 8-15 In this illustration, a flow control orifice has been added to the line from the directional control valve to the extend side of the piston. This will limit how quickly the 32,000-pound mass forces the piston into the full retract position by limiting flow out of the extend side of the cylinder. During the extend function, when the directional control valve is in the cross flow position, the smaller of the two orifices, the one with the smaller flow rate, will control how fast the piston can extend. The one on the extend line will limit the rate at which the fluid flows into the extend side of the cylinder, but the one on the retract side of the cylinder will limit how fast the fluid can flow out of the retract side of the piston. Since fluid cannot flow into the extend side until fluid flows out of the retract side, the one with the lower flow rate will control the speed.

Flow through an orifice to slow descent

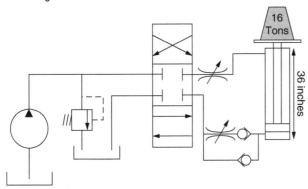

Figure 8-16 Here the orifices have been replaced with flow control valves. A flow control valve is an adjustable orifice. When the directional control valve is in the cross flow position, fluid flows through the lower check valve in the diagram to the extend side of the piston in the cylinder. The speed at which the piston will extend will be determined by the flow control valve in the line between the directional control valve and the retract side of the cylinder. When the directional control valve is moved to the straight flow position, fluid will flow through the upper flow control valve and into the extend side of the cylinder. That fluid flow, along with gravity, will start the descent of the piston within the cylinder. The speed at which the piston moves will be determined by setting the flow control valve located to the left of the upper check valve in the diagram.

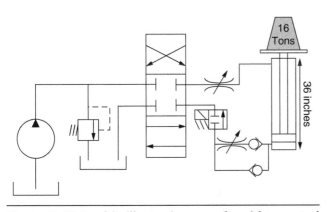

Figure 8-17 In this illustration, a **solenoid-operated valve** has been added to the extend side of the circuit. In order for the piston to extend the electric solenoid of the valve, it must be energized. Once energized and once the directional control valve is moved across the flow position, fluid can flow through the solenoid-operated valve and then through the lower check valve to the extend side of the piston. Fluid will then flow into the extend side of the piston, pushing the piston upward, and so fluid will move from the retract side of the piston through the upper flow control valve and return to the reservoir. When a solenoid-operated valve is turned off, the valve is

returned to a closed position. At this point, regardless of the position of the flow control valve, the piston will not move. The piston is therefore locked in place and the 32,000-pound weight will not force the piston to move. When the directional control valve is moved to the straight flow position and the solenoid-operated valves are activated, fluid flows into the retract side of the piston through the upper check valve, then through the lower flow control valve, then through the solenoid-operated valves, and afterward it returns to the reservoir, thereby allowing the piston to move to the retract position.

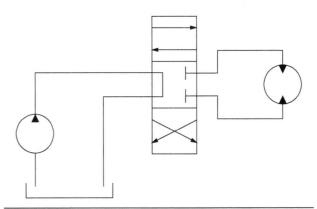

Figure 8-18 This is a simple motor circuit. When the directional control valve is in the center, or idle, position, fluid from the pump will flow to the directional control valve and immediately back to the reservoir. Because the outbound side of the directional control valve center position is closed, the motor is locked in place. Even though there might be a torque load on the motor, the motor should not rotate. When the directional control valve is moved to the straight flow position, fluid will flow through the upper host to the motor, then through the motor and back through the return line to the reservoir. When the directional control valve is moved to the cross flow position, the direction of the motor will be reversed.

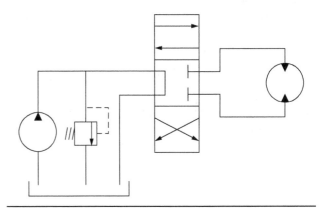

Figure 8-19 In this illustration, a pressure relief valve has been added. If the mass of the motor being asked to move these is too great—for instance, large enough to damage the system—the

pressure sensed by the pilot on the relief valve will cause the relief valve to open and the excess pressure will be fed back to the reservoir. When the directional control valve is in the center position, the open center on the supply side of the valve allows fluid to flow from the pump to the directional control valve, and then immediately back to the reservoir. Since there is little or no resistance to flow, the pressure on the supply side of the system will be 0 (or close to 0) psi. When the directional control valve is moved to a straight flow position, fluid will flow from the pump past the line to the pressure relief valve and onto the motor. If the load on the motor is excessive, pressure will build up to the point where the pressure relief valve opens and pressure will be reduced. In a cross flow position, the directional control valve direction of the motor is reversed and the pressure relief valve performs the same function.

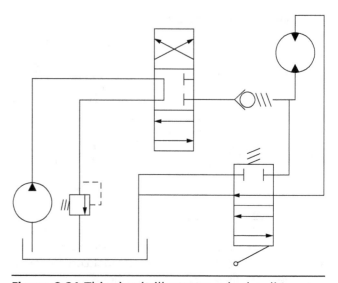

Figure 8-21 This circuit illustrates a hydraulic motor designed to raise a mass with a pulley and hold that mass in a fixed position until a manually operated valve is open to release trapped hydraulic fluid and allow the weight to be lowered. This simple design assumes that the integrity of the motor is high enough that the internal bypassing would be so low that the motor can stay in a fixed position. If there is an excessive amount of internal bypassing, in fact if there is any internal bypassing, then the motor would lower the mass regardless of the additional valving. The weight would be raised by moving the directional control valve to the cross flow position, fluid from the pump would then pass through the directional control valve, the check valve, through the motor, through the manual valve, and back to the reservoir. When the directional control valve is centered, fluid from the pump will simply pass through the directional control valve and return to the reservoir. On the outbound side of the directional control valve, the check valve prevents the motor from rotating as a result of gravity acting on the mass and causing the motors that pull it to rotate. Fluid flows are also blocked by the manually operated valve. When a manually operated valve is moved to the straight flow position, fluid is released between the motor and the check valve, and gravity can pull the mass down, thereby rotating the motor backward. A significant flaw in the design is that while the motor is rotating backward it can draw fluid from the reservoir. This could increase the possibility of variation and air entrainment in the hydraulic fluid.

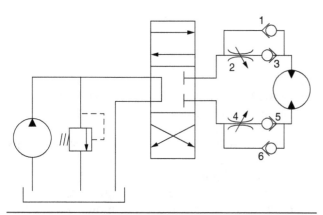

Figure 8-20 In this example, flow control valves have been added to limit the rotational speed of the motor. In the straight flow position, fluid passes through check valve 1 and onto the inlet side of the bidirectional motor. The fluid exits the motor, passes through check valve 5, then flow control valve 4, back through the correctional control valve, and on to the reservoir. Flow control valve 4 therefore controls the speed of the motor in the straight flow position of the directional control valve. When the directional control valve is moved to the cross flow position, fluid will pass through the directional control valve through check valve 6, through the motor, then through check valve 3, through flow control valve 2, back through the directional control valve, and then on to the reservoir.

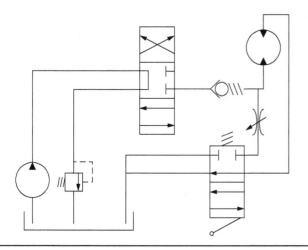

Figure 8-22 Here a flow control valve has been added between the motor and the manually operated valve. The flow control valve can be adjusted to alter the rate at which gravity will rotate the pulley and allow the mass to be lowered.

Brake

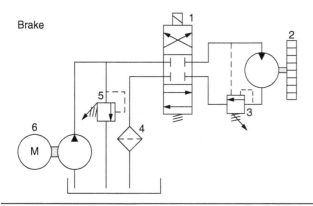

Figure 8-23 This diagram is more complete than the others seen in this chapter up to this point. The hydraulic pump is being powered by number 6, which is an electric motor. The fluid from the pump flows through a closed center directional control valve. In the center position, pressure will build up behind the directional control valve, eventually causing the adjustable pressure relief valve (5) to open, relieving pressure back to the reservoir. This reduces the load on the pump when the directional control valve isn't in the closed center position. The directional control valve is solenoid-operated in one direction only (1) and is spring-returned in the opposite direction. When the solenoid is activated, the fluid flows through the straight flow portion of the valve and continues onto the motor. The brake

valve (3) will remain closed until the pressure at either of the pilot lines increases above the setpoint for this pilot-operated valve. The line going to the upper supply for the pump senses when the pump has been activated because pressure will increase to the motor. When the pressure gets high enough to brake, the valve will open and allow the hydraulic motor to rotate. The lower pilot line is used when the hydraulic pump is off and there is a load resting on the pulley (2). If the load is excessive and will cause the pressure being generated by the motor in its rest position to open the brake valve, allow the load to be lowered, thereby saving the hydraulic motor from potential damage. Also note that a soldier has been added in the return line to the reservoir.

Brake

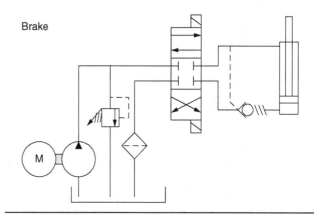

Figure 8-24 This illustration shows a brake circuit for a cylinder. When the directional control valve is in the straight flow position, flow from the pump will create a pressure in the retract side of the cylinder. As the pressure builds, it will eventually get to the point where the pilot-operated check valve will open and allow fluid to flow out of the extend side of the cylinder. If the electric motor is shut off, so that the hydraulic pump is not supplying flow, then the mass attached to the rod of the cylinder will apply a pressure downward, forcing the check valve closed and effectively parking, or setting the brake for, the cylinder. Please note that the directional control valve has a solenoid on each end. Also note that the directional control valve does not have springs to return it to the center position, as neither of the solenoids is energized. Therefore, the directional control valve will stay in the position that the last activated solenoid placed it in until the other solenoid is energized.

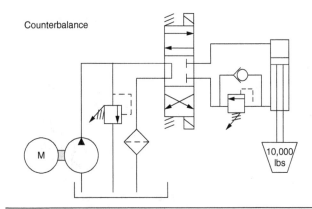

Figure 8-25 One of the most important circuits in fluid power systems (such as cranes and personnel lifts) is the counterbalance circuit. In this illustration, the retract side of the cylinder is used to raise a 10,000-pound mass. When the directional control valve is moved to the cross flow position, fluid flows from the pump through the directional control valve, through the check valve, and into the retract side of the cylinder. The piston then lifts the 10,000-pound weight. When the directional control valve is moved to either the center position or the straight flow position, two 10,000-pound weights attempt to push fluid back to the reservoir. Confounding this is a check valve and a pressure relief valve, which is being used as the counterbalance valve. To lower the weight it would be necessary for pressure to be built up between the pump and the extend side of the piston, this would increase the pressure on the retract side of the piston. When the pressure on the retract side of the piston meets or exceeds the pilot pressure for the counterbalance valve, the counterbalance valve will open and allow fluid to return to the reservoir through the filter. Note that the directional control valve is solenoid-operated in both directions and is spring-returned to center for both directions. The directional control valve illustrated here is the most common design, regardless of whether the valve is tandem, open center, or closed center.

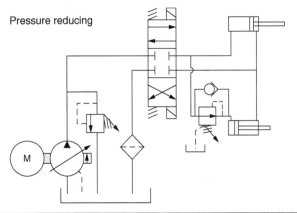

Figure 8-26 More complexity is added when fluid power circuits are designed to do more complex

tasks. A pressure reducing function is illustrated here. When an electric motor's power causes the variable displacement pressure-compensated hydraulic pump to rotate, the piston will move to full extension immediately. The fluid from the retract side of cylinder A will return to the reservoir through the filter. As pressure builds between the pump and the piston and cylinder A, fluid will begin to float the cylinder through the normally open sequence file (C). This can begin to move as the extend side pressurizes. Eventually the pressure on the extend side of piston B will be high enough to pilot the normally open sequence valve to the closed position. At that point, fluid can no longer flow into the extend side of cylinder B, and the force exerted by the piston in cylinder B is thereby limited. Notice the pump and the pressure sequence valve both have drains back to the reservoir so excess fluid can return to it, thereby limiting pressure.

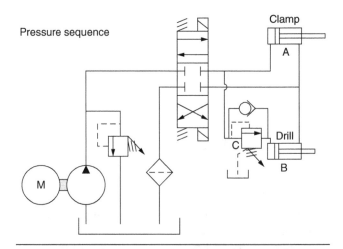

Figure 8-27 When two cylinders are in parallel, the one with the largest piston and/or the lightest load will move first. In this case, the cylinder that moves the drill is isolated from the pump by a pressure relief valve that is being used as a sequence valve (valve C). When the directional control valve is moved to the straight through flow position, fluid from the pump moves to the clamp cylinder (cylinder A) and extends the piston. Pressures will remain relatively low until the clamp has attached itself to whatever it is attempting to grasp. At that point, the cylinder will no longer move, pressure will build up, and at the preset pressure the sequencing valve will open and allow fluid to flow to the drill cylinder (B)—and so the drill will begin to move forward. When the flow control valve is moved to the cross flow position, fluid will go into the retract side of both cylinders. The piston and cylinder A will move freely through the return line while the fluid from the extend side of cylinder B will return to the check valve.

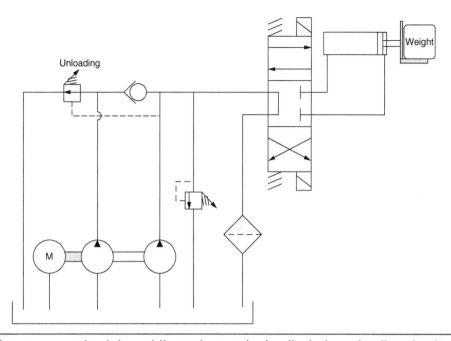

Figure 8-28 Another common circuit in mobile equipment hydraulics is the unloading circuit. In this circuit, an electric motor or perhaps a diesel engine is powering two pumps. One pump is large and uses a great deal of power from the motor. The other pump is smaller, has a lower flow capacity, and has a lower maximum operating pressure. The smaller pump, of course, utilizes less power from the motor. When the flow control valve is moved to the straight through flow position, the small pump supplies fluid to the extend side of the cylinder, attempting to make the piston move out. Meanwhile, the larger pump is drawing very little power from the motor because the fluid is passing through a normally open valve and returning to the reservoir. Note that the check valve is keeping the fluid from the small pump from returning to the reservoir. As the small pump is loaded, the pressure begins to increase, and at a predetermined pressure the pilot line that runs from the small pump out to the normally open unloading valve causes the unloading valve to pilot close. At this point, the fluid from the large pump is forced through the check valve and over to the cylinder to assist the small pump in extending the cylinder. When the directional control valve is in the cross flow position, fluid from the small pump will attempt to retract. If the act of retracting drives the pressure too high, then the pilot line will begin unloading and the large pump will once again assist the smaller.

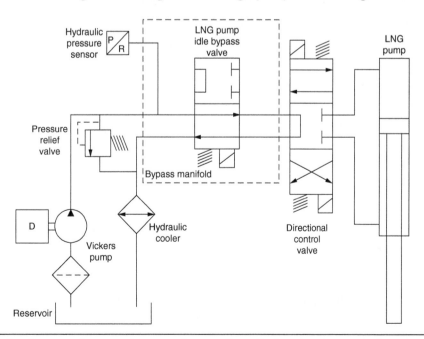

Figure 8-29 Hydraulic diagram for a system used on a liquefied natural gas engine to compress liquefied natural gas into compressed natural gas.

Review Questions

All of the review questions relate to **Figure 8-29**.

1. What type of power source operates the Vickers pump in this hydraulic system?

   A. A diesel engine

   B. A pneumatic motor

   C. A gasoline engine

   D. An electric motor

2. A technician is heard to say that the filter is located in a usual place on this system. This is because the filter is located:

   A. In the return line.

   B. In the bypass manifold assembly.

   C. In the supply line of the pump.

   D. In the cylinder return line.

3. The dashed line around the directional control valve indicates:

   A. There are several components inside of one assembly.

   B. The directional control valve is pilot operated.

   C. These components are optional.

   D. Nothing.

4. Instead of a filter in the return line, the location is occupied by a component called a hydraulic cooler. The purpose of this component is to reduce the temperature of the hydraulic fluid before it returns to the reservoir. In some cases, a hydraulic heater would be required. How would a symbol for a hydraulic heater be different?

   A. The double ended arrow would be pointing vertically.

   B. The double ended arrow would be inside a circle.

   C. The double ended arrow would be dashed.

   D. The arrow heads would be pointing inward.

5. The LNG pump idle bypass valve in the drawing is currently being shown in the:

   A. Straight through flow position.

   B. Center position.

   C. Cross flow position.

   D. Oscillator position.

6. How is the directional control valve operated?

   A. It is solenoid operated.

   B. It is solenoid operated with spring-return to center

   C. It is pneumatically operated.

   D. It is spring loaded with the solenoid returning to the center.

7. What is the purpose of the pressure relief valve?

   A. To regulate pressure to the cylinder

   B. To regulate pressure to the hydraulic cooler

   C. To route high temperature fluid to the hydraulic cooler

   D. To prevent damage to the system due to overpressurization

8. One advantage of a hydraulic system is that it is able to increase the amount of power from the power source—in this case, the diesel engine—so there will be a greater amount of power available to do the work than the diesel engine alone can provide.

   A. True

   B. False

9. The directional control valve is:

    A. Open center

    B. Closed center

    C. Tandem

    D. Bidirectional

10. When the directional control valve that supplies hydraulic fluid to the "LNG pump" is in the center position, the pump is:

    A. Able to move freely back and forth.

    B. Locked in place.

    C. Still moving.

    D. None of the above

CHAPTER

9 Electrical Controls

Learning Objectives

Upon completion and review of this chapter, the student should be able to:

- Describe the operation of electronic sensors and actuators in a fluid power system.
- Determine if the output voltage of the electronic sensors is in the proper range.
- Describe how data are communicated between the various control modules on modern trucks.

Cautions for This Chapter

- When working with an electrical system, sparks are always a possibility. Always make sure electrical circuits are turned off or powered down before connecting or disconnecting any component.
- Modern electronically controlled systems utilize microprocessors that can be damaged by static electricity. The human body is a major storage vessel for static electric voltage. Use extreme caution and/or a static wrist strap when working with electronic components.

Key Terms

amps	microprocessor	relay
analog	multiplex	resistance
capacitance sensor	negative temperature coefficient	rheostat
data bus	ohms	thermistor
Delta P	potentiometer	volts
J1587	pressure sensor	watts
J1939	reference voltage (VREF)	

INTRODUCTION

Like most technologies in the modern world, fluid power systems are quickly becoming dominated by electrical and electronic controls. Many seasoned veterans in the industry are frustrated by, and even fearful of, these control systems. Much of that attitude and concern is the result of an evolving industry and the fear of having to suddenly learn a new technology encroaching on the field that has been relatively self-contained for many decades. Yet there is little doubt that in the same way electronics has revolutionized the

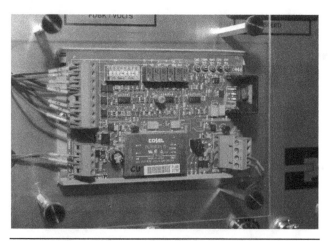

Figure 9-1 Over the past two decades, electronics has greatly simplified fluid power and hydraulic systems. That which would take several pounds of iron and very complex plumbing can be done with a couple of solenoids and some programmable integrated circuits. Pressure control functions, valve timing, and functions based on command inputs are much easier to do with programmable electronics than they ever were with heavy iron hydraulic system components. This does not mean that electronics-based systems can do without hydraulic valves, it simply means that the number of them can be reduced. In most cases, pressure relief and control of excess pressure is still being handled by classic mechanical hydraulic components.

Figure 9-2 Although effective and not dependent on the vehicle's electrical system to operate, complete manual controls tend to be heavy and bulky. Modern trucks are equipped with diesel oxidizing catalysts, diesel particulate filters, selective catalyst reduction systems, and SCR fluid tanks. These EPA-imposed components take up a lot of space on the truck and deprive the truck of valuable real estate for the addition of fluid power and other systems. Electronic control systems are far more compact and therefore do not compete as much for the real estate along the frame rails of the truck.

control and efficiency of diesel engines, electronics will and already has begun to revolutionize the control and efficiency of fluid power systems **(Figure 9-1)**. This chapter is in no way an effort to teach the entire technology of electrical and electronic flow power controls. The intent is only to refresh and supplement knowledge and skills acquired in previous electrical training or experience.

ELECTRICAL THEORY RELATED TO CONTROL CIRCUITS

Three general categories of components are used in the control of flow power systems. Sensor devices gather information about the operating characteristics of the system or about the characteristics of the environment in which the system is operating. These devices gather information about temperature, pressures, flow rates, exposure to light, exposure to humidity, and many other characteristics. The second category of electrical or electronic components is the control module. Some control models are simply a **relay** box wired with common, relatively large gauge, wire. In most modern fluid power applications, the control

modules are solid-state electronic devices utilizing transistors and **microprocessors**. The third category is the actuator. In most cases, a fluid power electrical actuator is a solenoid that changes the position of a valve **(Figure 9-2)**.

What Is Electricity?

Electricity can be explained many different ways. The physicist would contend that electricity is the movement of electrons from a point of surplus referred to as the negative terminal of the power source to the positive terminal of the power source. Benjamin Franklin would have probably said that electricity is an interesting phenomenon to study but probably has little or no practical value in daily life. To the typical consumer, electricity is a pervasive substance readily available in daily life **(Figure 9-3)**. It is immensely valuable for lighting, cooking, entertainment, and is generally taken for granted. To the technician, electricity is often thought of as a mysterious and troublesome necessity. To the fluid power technician, electricity and the troubleshooting of electrical/electronic control systems should be relatively easy. Electricity follows a similar set of rules to fluid power systems. Fluid power systems operate on the basis of pressure, volume, and force.

Figure 9-3 Electrical and electronic devices are ubiquitous in today's world. These electronics cook our food, brew our coffee, light our rooms at night, shave our faces, help us keep in touch with our friends and business associates, and keep track of our finances. This book was even written on an electronic device. In spite of the ever-present nature of electronics in the world around us, many people remain confused or at least intimidated by these devices. It is only natural that fluid power systems also be controlled by electronics—and it is also only natural that technicians are nervous about working with them.

Likewise, electrical/electronic control systems operate on the basis of pressure, volume, and flow as well. In the case of electrical systems, the pressure is called **volts**, the volume is called **amps**, and the force is called **watts**.

VOLTS

Since electricity is the flow of electrons through a conductor from the point of an electron surplus to the point of an electron shortage, there must be a force or pressure causing that flow. Just like in a hydraulic system where the fluid will flow from a high-pressure area to one of low pressure, similar occurs in an electrical system. With electricity, however, the equivalent of pressure is voltage. Voltage is the amount of push that moves the electrons through the conductor. Voltage by itself can do no work, cannot actuate a device, or even provide information to a controller about operating conditions or the environment.

Depending on the fluid power system being serviced, the technician can expect to work with both alternating current and direct current. Alternating current pushes the electrons out in surges or pulses whereas direct current provides a continuous flow of current **(Figure 9-4)**.

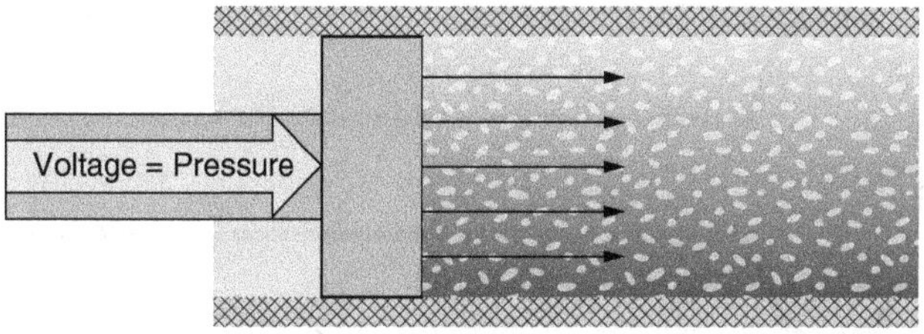

Figure 9-4 Voltage is electrical pressure. It supplies the push that allows current to flow through an electrical system in much the same way that pressure provides the push that allows hydraulic fluid to flow through a fluid power system.

Three primary voltage ranges come into play in a typical fluid power system. Stationary electric motors used to drive pumps are usually powered by high voltage in the 120 V to the 440 V range. This voltage source is supplied as alternating current. Technicians working around this level of voltage must use extreme caution, since under the right conditions even 120 V can deliver a fatal shock. As the voltage increases to 440 V, the possibility of serious injury or fatality increases. The second level of voltage is 12 V to 24 V. This is the voltage usually used to drive both electric motors and actuators or solenoids, as well as mobile fluid power applications, such as those on trucks or pieces of heavy equipment. Although most of the systems operate at 12 V, European and selected North American applications utilize 24 V. Because the 24 V system requires less amperage to do the same wiring work, the electrical components can be smaller and use lighter wiring to do the same amount of work. The third level of voltage is either 5 V or 8 V. This range is not used to do work since the current flow required would have to be extremely high, and therefore the wires would need to be very large. Thus, this voltage level is reserved for sensing.

AMPS

Amps are a measure of the flow rate of electrons, and thus the electricity flowing through an electrical circuit. In fluid power, the equivalent measurement would be gallons per minute. Specifically, one amp being pushed by one volt is the equivalent of 6.28 billion electrons passing a single point as the current flows through the electrical circuit. Just like how the flow rate of a single path fluid power system is the same at all points along the single path, in an electrical circuit the number of amps does not vary. And similar to how the flow rate in a hydraulic system is determined by the point of maximum restriction in that system, the number of amps passing through a circuit is determined by how much restriction to flow there is.

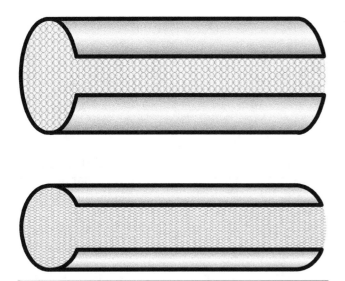

Figure 9-5 Electrical current flows through wire in much the same way that hydraulic fluid flows through the hoses of a fluid power system. The smaller the hose, the more restriction there is to flow in a hydraulic system. In an electrical system, the smaller the diameter of the wire, the more restriction there is to current flow. One of the most common mistakes made in installing or repairing a fluid power system on mobile equipment is to use wires with too small a diameter to connect the various electrical components together. Using a small-diameter wire will increase electrical resistance, increase the amount of heat in the wires, and likely impede the proper functioning of the fluid power system.

OHMS

The restriction to current flow and a circuit is called **resistance**. Resistance is measured in **ohms**. As a resistance of a circuit increases, the amount of current that can flow through that circuit will decrease. Many things can affect the resistance in a circuit. Among these is the material that the conductor, or wire, is made of, the length of the conductor, and the diameter of the conductor (**Figure 9-5**). The flow of the current against the resistance of a circuit can generate heat under many circumstances. As the temperature of the conductor increases, the resistance of the conductor also increases, and as the temperature of the conductor increases, the resistance also increases, which in turn decreases the amount of current flowing through the circuit (**Figure 9-6**). Another source of resistance is loose electrical connections, dirt, oil, grease, and corrosion.

WATTS

The amount of work that can be done by an electrical circuit is determined by the amount of voltage available

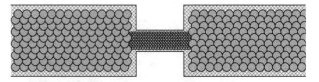

Figure 9-6 Ohms are a measurement of the conductor's resistance to flow. As the resistance in a circuit increases, the ability of current to flow decreases, and the amount of voltage available to operate a component also decreases. This is because resistance "uses up" voltage, which renders it unavailable for use in making the component operate. Resistance can come in the form of corrosion, using too thin a wire, having a frayed wire, or having a loose connection.

Figure 9-7 Watts are a measure of electrical energy doing work. It is directly translatable to horsepower using a mathematical formula. For most people, their contact with the term watts is limited to purchasing or replacing light bulbs. However, light bulbs are a perfect example of watts in action. Everyone knows the significant amount of heat that is given off by burning light bulb. This heat is work in action. Everyone also knows that as the wattage of the bulb increases, the amount of heat it gives off also increases.

to the circuit and the amount of current flowing through the circuit (**Figure 9-7**). A measurement of **volts** times amps is called **watts**. A watt is a measurement of power similar to horsepower. There are 746 W in 1 HP. If a fluid power system requires 10 HP to work, than 7,460 W would be needed to run an electric motor powering the hydraulic pump for that system. This would mean an electric motor running at 120 V would require over 62 amps to generate enough power through the hydraulic pump to make the system perform its assigned task. This rough calculation assumes that there is no energy loss in the transmission of that power from the electric motor to the hydraulic pump.

COMPONENT OPERATION

In a fluid power system, the primary job of the electrical or electronic system is to control pump speed, motor speed, and direction of flow. These controls may be used in a fluid power system to control the system, protect the system, improve the efficiency of the system, or prevent the system from damaging associated components of other systems. It should be noted that technicians will rarely encounter a system that has been modified with electronics just for the sake of having electronics. The primary driving force in a modern vehicle to add electronics to fluid power systems is the desire to integrate the operation of the fluid power system with other systems on the vehicle. A good example of this is the hybrid diesel—electric-powered utility trucks offered by several manufacturers. The electric motor designed to launch the vehicle and assist in driving the vehicle down the road is used to power the power takeoff (PTO) pump that operates hydraulic systems on the vehicle. An example of this would be a personnel lift truck, commonly referred to as a bucket truck. The hybrid electric motor powers the PTO, while the worker—for instance, a telephone lineman—is elevated in the personnel left. This allows hydraulic power to be present for the operation of the lift bucket without the engine having to idle the whole time the lineman is in the air. Many of these lift buckets are operated with a joystick. To increase the life of the battery for the hybrid motor once the lift bucket is in a fixed position, the hybrid motor shuts down. When the lineman moves the joystick, the hydraulic system electronics communicates with the hybrid system electronics and the electric hybrid motor starts up to power the PTO, which in turn powers the fluid power system, which in turn allows the lineman to be lowered or repositioned.

Sensors and Switches

Let us begin by looking at the lowest voltage portion of the system. This would be the sensors and switches used to gather data and control the operation of the fluid power system **(Figure 9-8)**.

TEMPERATURE SENSING

Temperature sensors can be placed in the reservoir or at other critical points within a fluid power system. Three common types of temperature measuring devices and circuits exist. The oldest and simplest is a

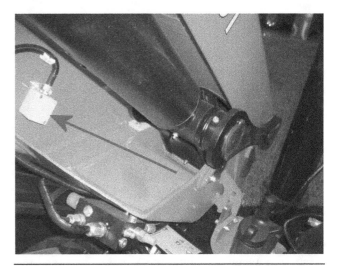

Figure 9-8 In many cases, the electronics found in fluid power systems is used not just for control but also for safety. The boom angle sensor (identified by the arrow) measures the angle of a bucket lift's boom. The lower the angle (the more parallel it is with the ground), the more likely the boom may tip over the vehicle upon which it is mounted. The boom angle sensor feeds information to a load limiting system, which in turn prevents the weight on the boom from tipping the vehicle over.

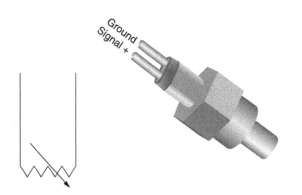

Figure 9-9 The typical temperature sensor has two wires and is generically referred to as a negative temperature coefficient thermistor. This term means it is a temperature-sensitive resistor whose resistance decreases as the temperature increases. One of the two wires is a ground and will always have 0 V on it. The other is the signal wire. This wire will typically carry between 0.5 V and 4.5 V when everything is operating normally. The higher the voltage, the lower the temperature. For this reason, a normal voltage on a temperature sensing circuit will usually be in the mid range around 2.5 V, give or take a volt.

thermocouple, the second is a positive temperature coefficient **thermistor**, and the third is the most common on modern equipment: the **negative temperature coefficient** thermistor **(Figure 9-9)**.

The thermocouple is the oldest of these devices. The internal structure of the thermocouple allows voltage to be generated when exposed to heat. The thermocouple is a very old and simple device. When it is exposed to an increasing temperature, the output voltage also increases; when it is exposed to a decreasing temperature, the output voltage decreases.

Thermistors cannot generate a voltage. They are resisters that will either increase or decrease their resistance as the temperature they are exposed to increases or decreases. The least common of the thermistor devices is the positive temperature coefficient (PTC) thermistor. The PTC thermistor changes resistance over a relatively narrow band. As the temperature to which the thermistors exposed increases, the resistance of the PTC thermistor also increases. The PTC thermistor circuit will typically have two wires: one wire carries a **reference voltage (VREF)** from the electronic control unit for the fluid power system to the thermistor; the other wire carries a voltage reduced by the resistance of the thermistor that is directly proportional to the temperature sent back to the electronic control unit. As the temperature that the thermistor is exposed to increases, the amount of voltage returning to the electronic control unit decreases **(Figure 9-10)**. The electronic control unit interprets the returning voltage as the change in temperature. In many cases, this temperature information can be broadcast by the electronic control unit on a **data bus** such as the J1939 data bus. The J1939 and other data buses will be thoroughly discussed later in this chapter. At this point, it is important to know that this temperature measured by the thermistor and electronic control unit for the fluid power system can be picked up and utilized by other electronic control systems on the vehicle.

The third and most common type of temperature sensing device is called the negative temperature coefficient (NTC) thermistor. Like the PTC thermistor, the NTC thermistor receives a reference voltage, typically 5 V, from the electronic control unit. The current being pushed by this voltage passes through the thermistor and returns along the ground wire to the ECU. The result of this design is that the ground wire always says 0 V on it regardless of the temperature, and the 5-V reference wire will have a voltage that varies inversely to the changes in temperature. That is to say, as the temperature increases, the voltage will decrease, and as the temperature decreases, the voltage will increase. The resistance value of PTC thermistors usually covers a very wide range. A typical PTC thermistor will have the resistance value well in excess of 100,000 ohms at a temperature of −40°C and have a resistance value well below 1,000 ohms at 100°C.

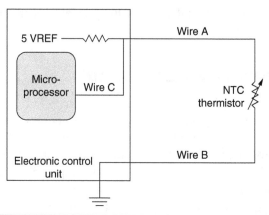

Figure 9-10 This is the layout of how a negative temperature coefficient thermistor is used to measure temperature associated with an electronic control unit. Inside the electronic control unit is a small voltage regulator that drops the voltage (usually 12 volts) down to 5 V. The 5-V supply then pushes a current through a protecting resistor (located inside an electronic control unit) and off to the thermistor. The internal protecting resistor has a fixed resistance. If the temperature that the thermistor is exposed to is such that the resistance of the thermistor is exactly the same as that of the fixed internal resistor, than the voltage between the two would be measured by wire C and a microprocessor and would be exactly 2.5 V. As the temperature rises, the resistance of the NTC thermistor would decrease, causing the voltage on wire A to also decrease. As the temperature decreases, the resistance of the NTC thermistor would increase and the voltage on wire A would also increase.

Most electronic control units have at least some self-diagnostic capability. The operator or service manual for each system should have diagnostic information, including retrieval and interpretation of diagnostic service codes. These diagnostic service codes result from the ECU recognizing a fault in one of its circuits. The codes come in two major categories: those that result from the ECU recognizing that the temperature being measured is outside an acceptable operating range; and those code sets when the ECU recognizes that the voltage returning from the sensor is either below or above the expected range of voltage. The first category of codes indicates that the fluid power system has a mechanical problem; the second category of codes indicates that there's a problem with electronics, and therefore the ECU is unable to monitor the condition of the mechanical system.

PRESSURE SENSING

It is the nature of fluid power systems that there are many places and opportunities to measure pressure.

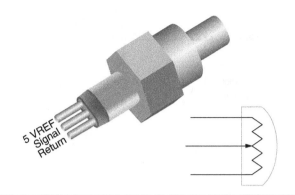

Figure 9-11 Several sensors used in the electronic control of fluid power systems have three wires. Typical of this would be a pressure sensor. A pressure sensor is a variable resistor device whose output signal increases as the pressure in a sampling increases. Typically, one wire will carry a 5-V reference (5 VREF) from the electronic control unit to the sensor. This final reference will be connected through the resistance element of the sensor to ground back at the electronic control unit through a wire normally labeled "return." The signal wire will carry a voltage that varies from 0.5 V at minimal pressure up to 4.5 V at maximum pressure **(Figure 9-12)**. The higher the pressure being sampled, the higher the output voltage of the sensor.

Obviously, there is the pressure of the fluid coming out of the pump, going to actuating cylinders, going to hydraulic motors, and even the pressure in the return line. Barometric pressure, which is the pressure of the air outside of the fluid power system, can also be measured to help the ECU determine how return fluid flow should behave, as well as how the fluid should behave when the pump attempts to move it out of the reservoir.

A typical pressure-sensing circuit will have three wires going to the sensor **(Figure 9-11)**. One wire carries the 5-V reference, a second wire provides a ground for the sensor, and the third wire carries a variable voltage from the sensor back to the ECU. In almost all cases, when the pressure is low, the output voltage of the sensor is also low; when the pressure is high, the output voltage of the sensor is also high. During normal operation of a pressure sensing circuit, the voltage on the signal wire from the sensor back to the ECU will never drop below 0.25 V and never go above 4.75 V. If a technician finds the voltage is outside the 0.25 V to 4.750 range, it can be assumed that there is an open short or ground either inside the sensor or in the wiring between the sensor and the ECU.

DELTA P SENSING

Pressure differential sensing is often referred to as **Delta P** sensing. Delta P sensing is a measurement

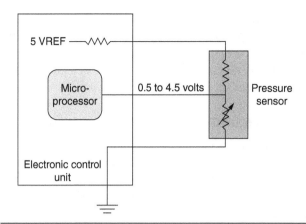

Figure 9-12 The 5-V reference destined for the pressure sensor originates in the electronic control unit. Like with the temperature sensor, the current destined for a pressure sensor also passes through an internal resistor. In this case, however, the internal resister usually has nothing to do with the operation of the circuit under normal conditions. The internal resister is there to protect the electronic control unit should the 5-V reference wire become grounded.

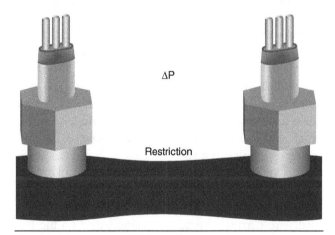

Figure 9-13 Bernoulli's law dictates that when a flow under pressure passes through a restriction, no matter how slight, there will be a drop in pressure on the outbound side. The term dP, or ΔP, is used to describe the function of measuring flow by means of placing a pressure sensor on either side of a restriction to flow. The greater the difference between the two pressure sensors, the greater the flow rate is interpreted to be.

of the difference in pressure between two points **(Figure 9-13)**. A typical place to measure the difference in pressure would be the difference in pressure between the inlet pressure at a hydraulic motor and the outlook pressure from the hydraulic motor. Although two sensors are required to make this measurement, the ECU subtracts one measurement from the other measurement. In most cases, a Delta P sensor will have three wires just like a standard **pressure sensor**. Each of the two pressures being measured is measured by separate sensors, the difference between the two pressures is detected internally by the sensor, and the output is sent on a single wire to the ECU.

FLOW SENSING

Flow can be measured in three ways. The first is through the use of a direct read device, the second is by measuring Delta P across a known restriction in the system, and the last method is by measuring a pressure drop, Delta P, across a known restriction in a bypass situated parallel to the main path of oil flow **(Figure 9-13)**. In a sense, all pressure measurement is pressure differential measurement: The pressure is either measured against a complete vacuum or against atmospheric pressure. However, any time a reference is made to Delta P, it specifically relates to a pressure measured between two points within the system.

POTENTIOMETERS AND RHEOSTATS

A general category of sensors is used to measure everything from angles and length to levels. These are **potentiometers** and **rheostats**. Both are variable resistors consisting of a resistive element, and a metal wiper moves across the resistive element. The rheostats will have two wires connected to them and will vary the resistance in a circuit. Moving the wiper will cause a change in resistance between the two wires, which will cause a change in current flow. A potentiometer is similar to a rheostat except that it has three wires. One wire carries a 5-V reference just like what is received by a pressure sensor. Also like a pressure sensor, a second wire is a ground for the sensor, and a third wire carries a variable voltage based on the position angle back to the ECU **(Figure 9-14)**. Similar to the pressure sensors, the normal range of operation is 0.25 V to 4.75 V. Outside of this range, the ECU will set a diagnostic code relating to open short or ground in the wiring to and from the potentiometer or in the potentiometer itself. The ECU is also programmed to recognize when the voltage is approaching the maximum normal range of operation. At that point, it can set a diagnostic code for excessive angle, overextension, low fluid level, or anything related to what the potentiometer is measuring.

LEVEL SENSING

Almost all fluid power systems have a method of measuring the hydraulic fluid level in the reservoir. In many cases, there is simply a sight glass or a sight tube mount on the reservoir. In systems where hydraulic

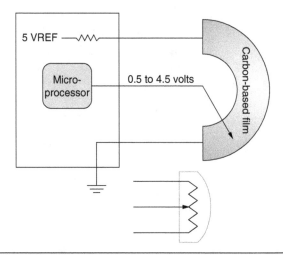

Figure 9-14 When position needs to be measured, the potentiometer is the most common component found. In its most rudimentary form, a potentiometer consists of a ceramic half-moon shape with a carbon-based film. Five volts will be fed into one end of the film, and the other end will be attached to ground. If one was connecting a voltmeter black lead to ground, placed the red lead on the carbon-based film closest to ground, and moved the red lead along the film toward the final reference, one would see that the voltage changes. The potentiometer has such a device. Contacting the carbon-based film is a strip of metal called a wiper. As the position changes, the wipers positioned inside the potentiometer also change as it moves along the carbon-based film; the output voltage of the sensor thus changes from a lower limit of 0.5 V up to a maximum limit of approximately 4.5 V. The microprocessor inside an electronic control unit senses this voltage and translates it into a position or a length, depending on the configuration of the equipment.

fluid level is recorded to a gauge or other instrumentation, a float and rheostat device is used. In this type of mechanism, a float rides on top of the fluid. As the fluid level changes, the float moves up and down with the surface of the fluid. The flow is attached to an arm, which in turn is connected to a rheostat that changes position as the arm connected to the float changes its angle **(Figure 9-15)**. The movement of the rheostat changes the resistance of the rheostat, which changes the resistance between the fluid global gauge and electrical ground. The result is that the angle of the needle on the gauge changes in direct correspondence to the fluid level. This type of sending unit can also be used to deliver a varying voltage to an ECU. If a reference voltage of, say, 5 V is delivered to the fluid level sending unit, then as the resistance in the rheostat changes, the resistance to ground for the 5-V reference will also change proportional to the fluid level.

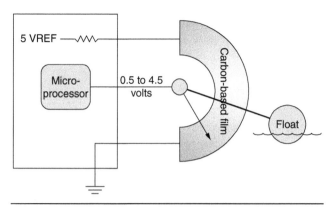

Figure 9-15 One example of how a potentiometer can be used is to attach the wiper to a float. If the float was placed in the hydraulic fluid reservoir, then as the level of the reservoir changed, the output voltage of this "level sensor" would change accordingly.

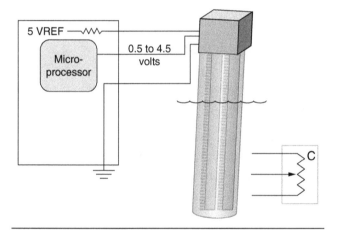

Figure 9-16 A relatively new category of sensors is used instead of a float mechanism to measure fluid level. These are called capacitance sensors. Here, two conductor rods or strips are submerged into the fluid. The sensor measures the resistance to the electrical charge between the two conductors to determine the fluid level. As the fluid rises, it effectively "shorts out" the conductor strips, thereby reducing the capacitance of the sensor. All the technician needs to be concerned with is that the output voltage will vary between 0.5 V and 4.5 V, depending on the fluid level. The documentation for the equipment will tell the technician whether 0.4 V represents an empty reservoir or one that is full. Typically, the lower voltage indicates a lower fluid level.

This information can then be interpreted by the ECU and translated into a fluid level.

In more sophisticated systems, the fluid level sensor is a **capacitance sensor (Figure 9-16)**. This capacitance sensor will have three wires, and like most pressure sensors, one of the wires will carry a

5-V reference, a second wire will be ground, and the third wire will carry a varying voltage between 0.25 V and 4.75 V back to the ECU. Also like the pressure sensor, if a voltage below 0.25 V or above 4.75 V is detected by the ECU, it can generate a diagnostic indicating there is an electrical problem (open, short, or ground) within the sensor, or within the wiring to and from the sensor. Another type of code can be generated if the fluid level is excessively high (the voltage approaching 4.75 V), or it is excessively low (the voltage approaching 0.25 V). This type of code indicates an actual problem with the fluid level itself.

LOAD LIFT SENSING

One of the most important benefits from the advent of electronic controls on fluid power systems is the ability to limit the lifting capacity of the system. This is especially of benefit to systems such as personnel lifts that are capable of extending their booms beyond the perimeter of the vehicle, or of the vehicle's outriggers. Extending the boom beyond the perimeter of the vehicle can make the vehicle unsteady. The farther it extends, or the more weight there is near the end of the boom, the more unsteady the vehicle can become. Eventually the vehicle will tip over and damage the vehicle or injure personnel.

Three primary factors determine a system's risk for tipping the vehicle over: the amount of load at the end of the boom; the angle of the boom over the perimeter of the vehicle; and how much the boom has been extended, should it have more than one element.

The first of these, weight on the end of the boom, is measured with a device referred to as a load moment sensor. Basically, this is a pressure sensor that determines how much force gravity is using to pull the lowdown. Putting this in simple terms, the load moment sensor is basically a scale much like a fish scale. The weight or mass is then reported back to the ECU as a varying voltage **(Figure 9-17)**. Typically, like with most sensors, the voltage varies from 0.25 V (with the lightest load or no load) to 4.75 V (at maximum load).

ANGLE SENSING

In most cases, angle sensing is done with a potentiometer. Sometimes there is a pendulum weight suspended from the wiper of the potentiometer so that as the angle changes with respect to gravity, the output voltage of the potentiometer also changes **(Figure 9-18)**. Other times, the body of the potentiometer will be connected to one side of a boom angle joint, and the wiper to the other side of the

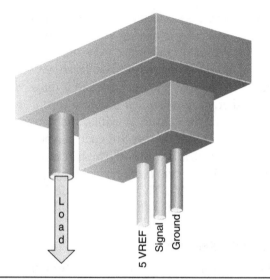

Figure 9-17 Most of the time, load in the system is determined by measuring the pressure in the system. As the load increases, the operating pressure in the system also increases. Sometimes, though, it is necessary to measure the load directly instead of using hydraulic pressure to find the information. Sensors used to measure load directly are often referred to as load moment indicators, or load moment sensors. There will be a physical attachment between the sensor and the mechanical component holding the load. Then, typically, there will be three wires: one carrying the 5-V reference (5 VREF); one carrying the ground for the 5-V reference; and one delivering the signal. As with most sensors, the output signal usually varies from 0.5 V (with small loads) to 4.5 V (with maximum).

angle joint. As the angle between the two sections of the boom changes, the output voltage of the potentiometer will change and report the angle back to the ECU.

EXTENSION

The traditional way to measure an extension is using a 25-turn potentiometer connected to the reel. The cable connects the reel and therefore the potentiometer to the end of the extendable boom. As the extendable boom lengthens, it pulls the cable, which rotates the reel, which then rotates the 25 turn potentiometer. As the potentiometer rotates, there is a change in output voltage **(Figure 9-19)**. Just like other potentiometers, this 25-turn potentiometer will have a 5-V reference wire running from the control unit to the sensor. There will be a ground wire with 0 V, and a single wire will carry a voltage ranging from 0.25 to 4.75 V.

BOOM ANGLE SENSING

Boom angle sensing is traditionally done by means of the potentiometer and a weighted pendulum. As the angle of the boom changes, the weighted pendulum remains parallel to the pull of gravity rotating the wiper inside of the potentiometer **(Figure 9-20)**. Once again, as with all potentiometers, there is a 5-V reference wire, a ground, and the signal wire. When the boom is horizontal to the ground, the output voltage of

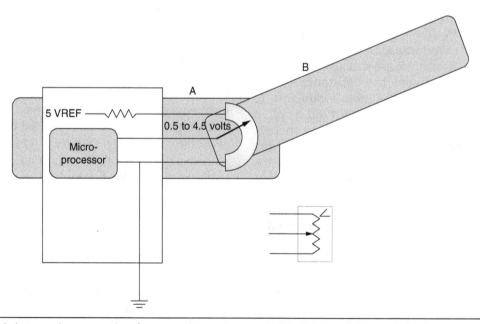

Figure 9-18 A joint angle sensor is often used to measure the angular difference between one section of a boom, or a materials handling device in another section. This is often necessary to calculate the amount of torsional load there is on the system, and thus the likelihood that the load will tip over the vehicle on which the system is mounted. This measurement is most commonly done with a simple potentiometer.

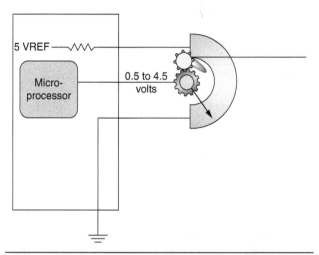

Figure 9-19 An extension—such as that of an extendable boom on a crane—is often also measured with a potentiometer. A cable is attached to the extending section of the boom while the sensor is attached to the non-extending section. As the boom extends, the cable spins a pulley, which rotates a gear, which in turn rotates the wiper of a potentiometer. As the boom extends, the cable rotates the pulley, which rotates the gear, which rotates the wiper. The voltage thus will vary through a typical range of about 0.5 V with the boom retracted, to about 4.5 V with the boom fully extended.

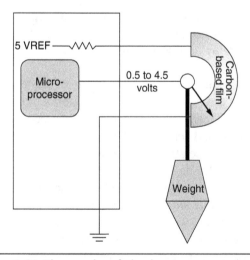

Figure 9-20 The angle of the boom on a crane can also be measured with the potentiometer. In this case, however, the potentiometer's wiper is moved by a pendulum. As the boom angle changes, the pendulum rotates the wiper and the output voltage from the sensor changes. On some systems, as the boom angle increases toward the vertical, the voltage will decrease, and on others it will increase.

the potentiometer is low. As the boom angle increases, the output voltage of the potentiometer increases from a little over 0.25 V to nearly 4.75 V when the boom is nearly vertical.

Electrohydraulic Actuators

The simplest electrohydraulic actuator in common use is a solenoid-operated control valve. These use a solenoid coil to create an electromagnetic field. Inside the coil is a free-moving armature, sometimes referred to as a soft iron core. The armature is spring-loaded away from the center of the magnetic field that will be generated by the coil of wire once energized **(Figure 9-21)**. The armature is in turn mechanically connected to the valve. In the case of a directional control valve, there is usually a solenoid at each end of the valve, one to pull the valve to the straight through flow position, the other to pull the valve to the cross flow position. Additionally, there are almost always springs to center the valve when neither of the two solenoids are energized.

A slightly more complex version is called a latching solenoid-operated valve. In this valve, the sliding spool remains in one of the flow positions, even though current is no longer flowing through the solenoid coil. These do not contain springs to re-center the spool valve. Instead, they often contain a small permanent magnet to keep the spool valve in the last position to which it was moved.

Many variations on electrohydraulic actuators are available. Most of them operate simply as described

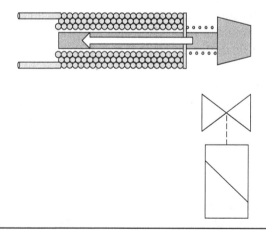

Figure 9-21 Oil flow control through the use of electronic components is quite common, and has been for many decades. A coil of wire and an iron core form a device known as a solenoid. If the solenoid is attached to a valve, then the opening and closing of the valve can be controlled electrically. The coil of wire forms an electromagnetic field. When energized, the electromagnetic field will attempt to center the iron core within the field. This causes the iron core to move, which in turn could open a valve allowing fluid to flow. In almost all cases, the solenoid-operated valves will be spring-loaded to the closed position. This does mean that sometimes it is spring-loaded to the open position.

earlier. Some of them are far more complex, utilizing internal logic circuits and internal sensors to perform a specific function. In the M1A1 battle tank used by the U.S. military, there is an electrohydraulic actuator that both receives commands from the operator and re-adjusts itself during operation. In this case, the gun operator sets the angle of the gun barrel in order to get a specific target. As the tank drives across the ground, and as the angle of the tank with respect to gravity changes, the servo automatically shuttles the spool valve back and forth to maintain the proper gun barrel angle with respect to gravity, and therefore its angle to the surface of the ground. This logic and move on the spool valve is all handled within the actuator.

Machine Control Using Relay Logic

Relay logic was the basis for electrical control of hydraulic systems for several decades. An electrical relay is a switch that is opened and closed by controlling a magnetic field, which is generated by a coil of wire located above or below the switch. In most cases, power is sent into the coil of the relay, and ground is provided by a switch that monitors either a parameter or a function. An example of this would be controlling the movement of a solenoid-operated valve based on temperature. A temperature switch would be located at the point where temperatures were to be monitored. The switch would be normally open. This means that when a current flows through the coil and generates a magnetic field, the switch would close. When current is not passing through the coil, the switch would be open. In this example, closing the switch would allow current to flow through the solenoid of the solenoid-operated valve, thus opening the valve and allowing fluid to flow through a cooling heat exchanger. When the temperature rises above the set point for the temperature switch, the switch will close. This energizes the electromagnetic coil inside the relay, which closes the relay switch, and which grounds the solenoid-operated valve. This opens the spool valve and allows a percentage of the fluid to flow through the cooling heat exchanger. Another relay could be incorporated that could have a normally closed switch. The coil on this relay could be used to open the relay switch if the temperature of the heat exchanger is too high. This would then form what is referred to in logic diagrams as an *AND gate*. With a circuit of this type, the temperature at the measured point would have to be above the set point for the switch, AND the temperature being measured of the heat exchanger would have to be below the set point in order for the solenoid-operated valve to open and allow fluid to go through the exchanger (**Figure 9-22**).

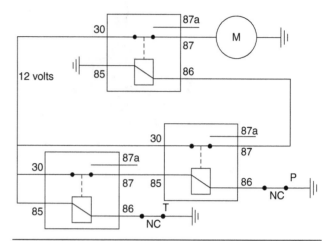

Figure 9-22 For several decades, the logic that controls the electrical aspect of fluid power systems has been done using relays. In this example, two switches labeled "T" for temperature and "P" for pressure must be closed in order for the electric motor to operate. Both switches are normally closed. If the temperature gets above a certain point, the switch would open, turning off the coil of the relay. Current would no longer flow from 85 to 86 and no magnetic field would be created by the coil. The switch would open. Instead of connecting 3 to 87, 30 would become connected to 87a. Current would no longer flow through the pressure relay's coil, the motor control relay would open and the motor would stop. If the pressure switch opens, then current will no longer flow to the coil of the motor control relay and again the motor will stop.

This is just one simple example of the cascading logic that can be formulated for the control of hydraulic systems. In many systems, temperatures, boom angles, extensions, and loads in all the routed-through relays control the operation of the hydraulic system. The problem with relays is that each control function will have a specific set point to either cause or inhibit operation. In other words, temperature would be measured by a switch and would either close or open the relay at the same temperature at all times, regardless of what is going on in the rest of the system. Relay logic can be a very effective method of controlling a hydraulic system that is limited by its inability to make decisions that vary based on environmental factors or what is going on the rest of the system.

Machine Control Using Integrated Circuit and Microprocessor Controls

The phrase "microprocessor control" seems rather intimidating, but in truth a microprocessor is little more than a box containing relays. The microprocessor has the ability to be programmed for the logical control

Figure 9-23 Over the last couple of decades, logic controls using relays have given way to the cheaper and more compact microprocessor. An additional advantage of the microprocessor is that its functions can be altered by simple reprogramming it instead of having to replace sets of relays and rewire those relays.

of the millions of relays that make it up. Modern trucks use microprocessor-controlled engines, transmissions, braking systems, and even starting systems **(Figure 9-23)**. Microprocessors used to control fluid power systems, allowing it to be fully integrated with the computer control system and the networking system of the truck.

One of the more stunning examples of this is the integration of a fluid power personnel boom in the new hybrid system by Eaton. These new hybrid-powered trucks use a diesel engine and an electric drive to move the vehicle down the road. When initially launching the vehicle from a stop, the electric motor gets the vehicle rolling and then the diesel engine begins to pick up the load and accelerate the vehicle up to cruising speed. This saves considerably on fuel consumption. However, the time when these personnel boom trucks burn the most fuel for the least amount of work being done is when the vehicle is sitting still and the engine is running to power the hydraulic boom. A hybrid truck set up to operate a fluid power system by means of the PTO, referred to as an ePTO, will not use the engine while the hydraulic system is being operated, but instead will use the hybrid electric motor.

Imagine the following scenario. A technician employed to change parking lot light bulbs arrives on the jobsite. Each lamp pole has four bulbs, and each one takes approximately 20 minutes to change. In a conventional system, the engine would have to be started and run the whole time the technician was in the

personnel lift bucket. This means that for each lamp pole the truck's engine would be running for 80 minutes. For each lamp pole attended to, the truck would spew the exhaust from a wasted gallon of gasoline or diesel fuel. With a microprocessor control fluid power system integrated into the data backbone of a hybrid truck, the engine can be shut off and an electric motor used to power the PTO pump. Since an electric motor can be controlled by the same microprocessor system that is integrated with the fluid power system, the hybrid electric motor can shut off when the personnel bucket is not moving and be powered up to operate the fluid power system when the personnel bucket needs to be moved.

J1587, J1939, AND THE CONTROLLER AREA NETWORK

Beginning in the early 1990s, the Society of Automotive Engineers established a method of communication between the onboard computers on medium and heavy-duty trucks **(Figure 9-24)**. The standard became known as **J1587**. J1587 is an SAE document that established the language and communication protocols for communication between a truck's onboard computers. In the early 1990s, a typical home desktop computer could communicate over a phone line with a bulletin board service (this was in the days before common use of the Internet) at 360 bits of information per second. The term for bits of information per second is the word *baud*. The J1587 communication rate was 9,600 baud. This was incredibly fast for its day. As trucks became equipped with more and more computer-operated and -monitored systems, it became necessary for there to be a faster communication protocol. This brought about **J1939**. This new communication protocol was designed to operate at 250,000 baud. With J1939, it became possible for body builders to integrate their systems with the systems manufactured into the truck. By connecting the fluid power computer to the vehicle data bus, it was now possible for instrumentation, lighting, engine operation, transmission operation, ABS operation, and virtually any system on the truck to be integrated with the operation the fluid power system.

PROGRAMMING

One of the greatest benefits of the microprocessor is its ability to be reprogrammed in order to perform new tasks or to compensate for design faults. The high baud rate of J1939 allows for the electronic control modules on the truck to talk to each other. It also allows each

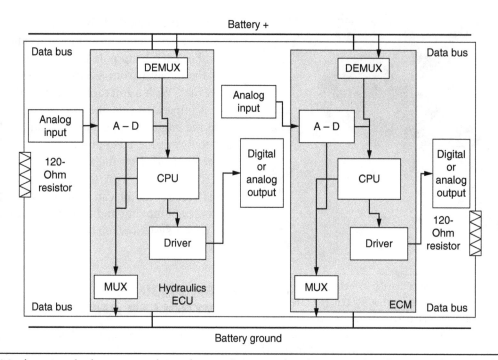

Figure 9-24 Modern trucks have complete electronic control unit networks. On a modern truck, the engine controller can communicate with the antilock brake controller, which can communicate with the transmission controller, which can communicate with the instrumentation controller, which can communicate with the lighting controller, which can communicate with the exhaust after-treatment controller, which in turn can communicate with the fluid power system controller. This complex communication system allows for increased safety and reduces the number of sensors required by each of the control units on the network. For instance, the ABS wheel sensors can report to the fluid power system controller to inform the fluid power system controller that the vehicle is moving. At that point, if appropriate, the fluid power system can shut down, retract outriggers, plant outriggers, or perform whatever is necessary to make the equipment safe to use. **Multiplexing** allows the electronic control modules that control the various systems of the truck to share information along a set of wires known as a data bus. **Analog** (variable voltage) information is input into one controller which processes the information into computer code. That code is then multiplexed (MUX) and broadcast along the data bus for another controller to receive. Once received the information is de-multiplexed (DEMUX) into the language used by that controller and the translated digital information is then used by the receiving controller.

of the electronic control units to communicate with a diagnostic tool, but also allows a diagnostic tool to alter the programming of the electronic control unit. The programmability of the microprocessor-controlled fluid power system is especially noteworthy when it comes to safety. For instance, it is possible to set specific safety requirements before the fluid power system can be allowed to engage the boom on a personnel lift truck. What follows are a few instances of how this could work:

- The ABS system monitors wheel speed. If wheel speed above zero is indicated by the wheel speed sensors, then the ABS controller can report this on J1939. The fluid power controller can respond to this information by preventing the PTO pump from engaging if there is any vehicle speed indicated.

- If the transmission fails to report that it is in neutral on J1939, then the fluid power controller can respond by not allowing it to engage.
- When the hydraulic system reports on J1939 that it is in need of hydraulic flow from the pump, the engine can respond by increasing the rpm to preprogram levels to provide the proper flow.

The flexibility of the microprocessor allows both the fluid power system as well as the engine, transmission, and ABS systems to work in conjunction with one another and to adapt themselves to the needs of the other controllers.

Microprocessor control systems use a variety of inputs such as the sensors already discussed in this chapter. In addition to these sensors providing information that affect the operation of the fluid power system, they can also provide telemetry to allow the system operator to monitor the operation of the system from another part of the vehicle, or even from the other side of the planet.

CONTROLLER DATA COMMUNICATION

On a given vehicle, there may be as many as a dozen electronic control units. Each one of the electronic control units is assigned a message identifier (MID). Each MID is assigned, under J1587, a specific three-digit number by the Society of Automotive Engineers. For example, the controller for the primary drive engine on the vehicle is assigned the number 128. The hydraulic system will be assigned a number based on the task it is going to perform. For instance:

112–127	Unassigned—Available For Body Builder
128	Engine #1
130	Transmission
131	Power Takeoff
136	Brakes, Power Unit
140	Instrument Cluster
146	Cab Climate Control
150	Suspension, Power Unit
159	Proximity Detector, Front
164	Multiplex
166	Tires, Power Unit
171	Driver Information Center
204	Vehicle Proximity, Right Side
205	Vehicle Proximity, Left Side
249	Body Controller

This list is made up of the MIDs most likely to affect, or be affected by, a fluid power system. MID 131 or MID 249 are the two most likely to represent the fluid power system.

Parameter Identifier/Subsystem Identifier

Each control module will be monitoring many input and output parameters and perhaps managing some subsystems as well. In a fluid power system, a subsystem might be pump controls, directional control valves, or perhaps actuator controls. Well over 1,000 parameters and subsystems can be monitored by vehicle electronic control units. Of course, only a small number of these relate to fluid power systems. Examples of ones that do relate to fluid power systems include:

PARAMETER IDENTIFIERS

PID   Parameter

Single Data Character Length Parameters

56	Transmission Range Switch Status
65	Service Brake Switch Status
70	Parking Brake Switch Status
71	Idle Shutdown Timer Status
89	Power Takeoff Status
90	PTO Oil Temperature
91	Percent Accelerator Pedal Position
93	Output Torque
124	Transmission Oil Level
125	Transmission Oil Level High/Low
126	Transmission Filter Differential Pressure
127	Transmission Oil Pressure

Double Data Character Length Parameters

134	Wheel Speed Sensor Status
160	Main Shaft Speed
161	Input Shaft Speed
162	Transmission Range Selected
163	Transmission Range Attained
186	Power Takeoff Speed
187	Power Takeoff Set Speed
188	Idle Engine Speed
189	Rated Engine Speed
190	Engine Speed
191	Transmission Output Shaft Speed
233	Unit Number (Power Unit)
234	Software Identification
235	Total Idle Hours
236	Total Idle Fuel Used
245	Total Vehicle Distance
246	Total Vehicle Hours
247	Total Engine Hours
248	Total PTO Hours
249	Total Engine Revolutions
251	Clock
252	Date
253	Elapsed Time

Single Data Character Length Parameters

Subsystem Identifications (SIDs) 1 to 150 are not common with other systems and are assigned by SAE. SIDs 151 to 255 are common among other systems and are assigned by the SAE Subcommittee.

Common SIDs

248	Proprietary Data Link
249	SAE J1922 Data Link

250   SAE J1708 (J1587) Data Link
251   Power Supply
252   Calibration Module
253   Calibration Memory

FAILURE MODE IDENTIFIERS

For each parameter, and for many subsystems, there is a relatively short set of code numbers that specify how that particular parameter, component, or subsystem has failed. These identifiers are designed to inform the technician about exactly what type of problem causes a parameter or subsystem diagnostic code to be set.

Failure Mode Identifiers (FMIs)

0 Data valid but above normal operational range
1 Data valid but below normal operational range
2 Data erratic, intermittent, or incorrect
3 Voltage above normal or shorted high
4 Voltage below normal or shorted low
5 Current below normal or open circuit
6 Current above normal or grounded circuit
7 Mechanical system not responding properly
8 Abnormal frequency, pulse width, or period
9 Abnormal update rate
10 Abnormal rate of change

11 Failure mode not identifiable
12 Bad intelligent device or component
13 Out of calibration
14 Special instructions
15 Reserved for future assignment by the SAE Subcommittee

An example of putting this all together would be a fluid power system that has its own dedicated diagnostic software that can be installed and used on a desktop or laptop computer. The vehicle comes in with an inoperative hydraulically operated personnel lift. The technician uses his laptop software to locate the source of the problem. The software retrieves an MID 249, and a PID 90 with an FMI of 0. This would indicate that the body controller—in other words, the controller for the fluid power system—has a problem with the hydraulic oil temperature sending unit. The FMI 0 indicates that the data are valid—therefore, the circuit is working, but the temperature is above the normal operational range. The technician would need to find a reason why the oil temperature for the fluid power system is excessively high. The very same MID and PID combination with an FMI of 4 would indicate that the problem is not oil temperature–related but instead the problem is in the temperature sensing circuit itself. To be more specific, it appears that the signal wire is shorted to voltage.

Summary

While at first microprocessor/computer control of the hydraulic system would seem to add a significant amount of complexity, the truth is it makes diagnosing and troubleshooting much easier. During the next few years, the industry will see a much greater integration about systems for trucks and equipment with their onboard network. At the very least this will reduce the mass of wires and optical fiber that has been seen in mobile fluid power systems of the past. System integration will undoubtedly reduce the complexity and weight of the systems while increasing their dependability and ease of use.

Review Questions

1. The equivalent of pressure in an electrical system is called:

   A. Ohms.   C. Watts.

   B. Volts.   D. Resistance.

2. All of the following can cause resistance in an electrical circuit except:

   A. Corrosion.   C. Too small of a wire.

   B. A loose connection.   D. Too large of a wire.

3. The device used to measure temperature and report the temperature to an electrical controller in a typical fluid power system is called:

   A. An NTC thermistor.   C. An FET transistor.

   B. A PTC thermistor.   D. A thermocouple.

4. For most sensors, output through the normal operational range varies between:

   A. 0 V and 5 V.   C. 0.5 and 0.45 V.

   B. 0 V and 12 V.   D. 0.5 and 4.5 V.

5. A ΔP sensor measures:

   A. Pressure.   C. Pressure differential.

   B. Electrical pressure.   D. Negative pressures.

6. When a capacitance style sensor is used in a fluid power system, it is most usually used to measure:

   A. Fluid level.   C. The angle between the flexing components on the truck.

   B. The angle from the ground.   D. The torsional load on the truck.

7. Potentiometers can be used to measure, either directly or indirectly, all but which of the following:

   A. Torsional load.   C. Temperature.

   B. Boom angle.   D. The angle to the ground.

8. In an electrical diagram illustrating a relay control for logic in a fluid power system, a rectangle with a diagonal line within the rectangle indicates:

   A. A coil of wire.   C. A switch.

   B. An electric motor.   D. A resistor.

9. Even though the microprocessor is far more compact and therefore takes up less real estate on the truck than a relay logic control, it has yet to catch on in the industry because of its enormous expense.

   A. True   B. False

10. The main advantage of multiplex control systems on a truck is their ability to:

    A. Allow different electronic control modules to share information.   C. Both "A" and "B".

                                                                              D. Neither "A" nor "B".

    B. Minimize the number of sensors installed on the truck.

CHAPTER

10 Maintenance

Learning Objectives

Upon completion and review of this chapter, the student should be able to:

- Describe the minimum process for doing routine inspections.
- In the absence of the manufacturer's recommendations, build a maintenance plan for any mobile equipment fluid power system.
- Perform an inspection.

Cautions for This Chapter

- Inspecting a fluid power system requires a technician to be very close to the components that are being inspected. Thus, it is absolutely essential that safety glasses, or even a full face shield, be worn during the inspection. It is also advised that the inspecting technician wear an oil-resistant apron, steel toe boots or shoes, and gloves.
- Be sure to follow the manufacturer's recommendations for maintenance intervals. Perform the tasks suggested at a minimum.

Key Terms

Bourdon tube   infrared   inspection

INTRODUCTION

Fluid power systems, like most other systems, require routine maintenance **(Figure 10-1)**. This includes verifying the condition of the oil (hydraulic fluid), measuring the restriction in the filtration system if possible, or at least confirming that the last change of the filtration media was within a reasonable time frame. A visual **inspection** for leaks, and damaged lines and hoses (if the system is equipped with a cooler, thus confirming the efficiency of the cooler) is also important. Electrical controls and wiring should be checked for loose connections, corrosion, damaged

wires, and proper operation. Although many electronically controlled systems are designed to monitor their own need for maintenance, a technician should never depend upon these systems over their own eyes, ears, hands, and handheld equipment.

DAILY MAINTENANCE CHECKS

Begin daily inspections with a complete walk around the system. Check the condition of the hydraulic fluid and its level. Manually activate the systems and observe a complete cycle of operation. Make sure that

Figure 10-1 All fluid power systems require routine maintenance. This is especially true of mobile equipment. Mobile equipment operators tend to be experts at either driving or using equipment on their truck to perform repair and maintenance on other equipment. This means that in most cases they are not experienced, or are experts, in hydraulic systems. When equipment does come in for service, the vehicle technician needs to at least inspect the condition of the fluid power system.

not only are there no hydraulic leaks, but also that the operation of equipment is smooth and its mechanical components operate smoothly and within the required specifications **(Figure 10-2)**. Almost all systems are equipped with some sort of warning device for when the system is functioning or operating in a dangerous manner. Manually test the warning devices to make sure the equipment and the operators will be safe. While the equipment is cycling, make sure all of the gauges indicate the correct operating pressures.

WEEKLY OR MONTHLY MAINTENANCE CHECKS

Depending on the duty cycle of equipment, a more thorough check is required on a weekly or monthly basis. Check the filters and clean or replace as necessary. In many cases in the hydraulic oil pickup in the reservoir, there will be strainers or screens that may need to be removed and cleaned. Restricted filters, strainers, and screens can reduce the ability of the

Figure 10-2 Even the simplest hydraulic equipment, such as this engine hoist, requires routine inspection and maintenance. This may seem like a simple machine that would not require routine service. The technician should keep in mind that this engine hoist may have as much as nearly 2,000 pounds hanging from it. The author has worked in many shops where it was necessary to continuously pump the handle in order to keep the engine suspended. This presents an extreme safety hazard in that the engine may suddenly drop while attempting to adjust the engine mount, leading to a crushed hand. The cylinder should be inspected for leakage, and the attachment point of the cylinder's rod to the horizontal boom should be inspected for wear and a snug fit. The mounting for the cylinder at the lower end of the vertical boom should be inspected for cracks or other damage. The cylinder should be inspected for signs of wear, leakage, or internal bypassing. Then, the critical step is to make the necessary repairs as indicated by the inspection.

pump to draw the hydraulic oil from the reservoir. This can cause a pump to cavitate, reducing the efficiency of flow throughout the system, and cause damage to the pump and possibly to other components in the system. Also, pressure can build up behind filters due to the restriction of flow through the filter. Eventually, the pressure can become great enough to tear the filter or screen, allowing contaminants ordinarily filtered to be passed through the system, or worse, allowing particles of the screen or filter to pass through the system.

Although it is important to check piping and hoses for broken lines, leaks, kinks, or blockages on a daily basis, it would be extremely beneficial to check these thoroughly on a weekly or monthly basis. Rust around welded pipe junctions can indicate tiny fractures around the welds that could lead to future failure. In many systems, the hydraulic hoses seem to be covered in a light film of oil. The oil can originate from minor seepage, or it could be the result of oil being spilled during a previous service or from the oil film that covers most roads. This oil can attract dust and dirt, making it difficult to tell if the hose is leaking. Use a mild detergent to wipe the oily film off the hoses and rinse the detergent away with plain water. Be sure to follow approved local ecological guidelines when cleaning the hoses or any other part of the fluid power system. The clean hoses and lines will make it easier to recognize new leaks during the daily inspections.

Many fluid power systems have heat exchangers to cool, or in some cases warm, the hydraulic fluid. On mobile fluid power systems, the coolers are usually liquid-to-air heat exchangers similar to the radiator in a car. Some of them have fans and some of them do not. The cooler should be inspected for dirt and for air flow restrictions. Some environments, such as coastal areas, can cause a coating to build up on fins and tubes of the cooler **(Figure 10-3)**. This coating can impede the ability of the fins to give up heat to the air passing across them. Even if the cooler looks clean, it should be thoroughly cleaned with water during each of the weekly or monthly services.

On mobile equipment, heaters are usually liquid-to-liquid heat exchangers. There is an assumption that the

Figure 10-3 A critical part of any maintenance inspection is to ensure that the heat exchangers are capable of properly cooling the hydraulic fluid. In many cases, air is forced through the heat using an electric fan. The proper speed, air flow, and operation of the electric fan should be checked on each inspection. Additionally, the air channels should be inspected for signs of contamination, blockage, or any damage that might impede airflow. And, of course, the heat exchanger and its connections should be checked for signs of leaks.

truck will be driven to a job site before the fluid power system is used. As a result, the engine coolant will be close to 100°C (212°F). When the fluid power pump is started, the hydraulic oil is moved through the heater when the heat from the coolant is transferred to the oil. Since there are no exposed fins or tubes to clean, inspecting for leaks and physical damage is all that can be done.

Quarterly or Annual Requirements

The first thing that separates the longer-interval services from the shorter-interval inspection is replacement or detailed analysis of the condition of the fluid. Each fluid power system manufacturer has recommended intervals for the replacement of the oil and the filters. In reality, these intervals are often the maximum service intervals. There is a lot that goes into determining recommendations for replacement of oil and filters that goes beyond the design requirements of the system. If a manufacturer recommends frequent replacement of fluids and filters, they quickly gain a reputation for building high-maintenance systems. Therefore, they tend to publish service intervals that are appropriate for operation of the equipment in ideal conditions and in an ideal environment. They usually have a footnote indicating that the services would need to be more frequent in hostile or rough service environment. Therefore, during the quarterly or annual inspections, a sample of the oil should be sent to a lab for analysis. The contents of the oil can tell the technician about the level of contamination and in many cases the extent of wear on the components of the system. Of course, on smaller systems it would be more cost-effective to simply change the oil and filters rather than doing the oil analysis.

Electrical connections should be inspected, taking more time and effort than during the shorter-interval inspections **(Figure 10-4)**. If electrical connections are held in place with a screw, bolt, or other removable fastener, the screw should be removed and the connection cleaned. All connections should be inspected for tightness.

Check the Pressure Gauges and Calibrate as Needed

Pressure gauges, especially the old **Bourdon tube** design, require occasional calibration. These gauges date back to 1849 when they were patented by Eugene Bourdon of France. It has been the primary gauge design used to measure pressure for over 160 years. Being a mechanical device, however, it is subject to

Figure 10-4 All electrical connections should be inspected for tightness, evidence of corrosion, physical damage to the electrical device, and physical damage to the connector. In many cases, such as in the solenoid, the electrical connectors are inside the assembly. Since it is not practical to take the assembly apart to inspect the connections, the technician should gently pull on and wiggle the wires for evidence that they might be loose inside the components. A few inches from this device there is an electrical connector attached to the harness of the fluid power system. This connection should be thoroughly checked to make certain there will be no resistance to current flow and thereby ensure proper operation of the controls for the fluid power system.

Figure 10-5 One device often overlooked during a maintenance inspection is the condition of the pressure gauges. Even with state-of-the-art electronics, the pressure gauge mounted on mobile equipment fluid power systems still tends to be the mechanical Bourdon tube pressure gauge. This has been the industry standard for 160 years. The problem with Bourdon tube gauges is that they can go out of adjustment due to changes in the characteristics of the tube itself, or as a result of damage or misalignment of the gear mechanism that moves the needle. This is the inside of a gauge that was misaligned due to a gear mechanism and whose reading was off by over 150 psi.

wear and deterioration **(Figure 10-5)**. Thus the accuracy of these gauges must be checked and they should be calibrated, if necessary, every year.

In more modern systems, pressure is monitored by strain gauges and displayed on digital panels. The design of strain gauges is such that they cannot be calibrated. This of course does not mean that their accuracy should not be checked on an annual basis.

Check the Operation of Warning Safety Devices

Larger fluid power systems, especially large stationary systems, will have safety devices and warning devices that activate when the system behaves improperly **(Figure 10-6)**. In mobile equipment applications, these devices prevent excessive weights from being lifted, prevent weights from being lifted at dangerous angles, or warn system operators of potential pressure and flow problems.

Systems such as cranes and personnel lifts are usually equipped with systems that prevent high-weight loads from being extended at an angle that could cause tipping of the truck. When the crane boom or bucket is vertical to the mounting point on the truck, all of the

weight is directed downward and is not contributing torque that could roll or tip the truck over. When the boom is directed at a 90-degree angle to the vertical, all of the weight of the boom and the bucket are contributing torque and therefore contributing to the possibility that the truck may roll and tip over. Although some of these systems can be extremely complicated for those not trained to work on them, their basic operation is actually quite simple. If there is a single arm to the device, there will be a potentiometer to measure the angle of the arm with respect to the ground. If the boom consists of a double arm, there will be two potentiometers to measure the angle of each, and the control module will calculate the total angle with respect to the ground from these two measurements **(Figure 10-7)**. A second type of sensing device, basically a pressure gauge, measures how much weight is either suspended in the bucket or hanging from the boom. The controller for the system uses the electronic input devices to calculate the amount of torque being applied to the truck based on the angle of the boom and the weight that is suspended from the boom. When loads on the boom are high, the controller will prevent the boom from being lowered to angles approaching parallelism with the ground. With lighter loads, the

Figure 10-6 Hydraulically operated safety switches should also be inspected for proper function. Although most fluid power systems have pressure relief valves and other safety mechanisms to ensure that the system is not overpressurized, thus leading to a catastrophic failure of the hoses and other components, electrical safety switches to shut off the pump or change the position of the directional control valve are also essential for safety. At least inspect them to make sure there is no evidence of physical damage or leakage. If the design of the system permits, it would also be a good idea to pressurize the portion of the system being monitored by the pressure switch to the point where the switch should cut off. This will verify that the safety function operates properly.

Figure 10-7 As mentioned in preceding chapters, other safety devices monitor angles and loads to ensure that torsional loads do not cause the vehicle to tip or tilt. In many cases, these components are controlled by an electronic control unit and have the ability to be monitored for proper operation through software on a laptop computer. Any system that uses this level of technology should be tested by connecting the software to the ECU and verifying that the system is working properly and that there are no diagnostic codes.

controller will allow the boom to be lowered to angles that approach parallelism to the ground.

In most cases, modern versions of these systems are based on digital controllers. Proper operation of these systems can be verified with the use of diagnostic software provided by the manufacturer. If the system is older, there are usually weight angle specifications that can be tested. A specific weight is placed on the end of the boom, either in the bucket or suspended from the boom, and the boom is lowered to an angle nearly parallel with the ground. When the boom achieves a specific angle, the controller will shut down the hydraulic system and prevent the boom from being lowered any further.

These systems are critical in preserving the safety of the operators utilizing the equipment. These safety checks should be done any time repairs are being made on the system, when the system is in for routine maintenance, and during annual, semiannual, or quarterly inspections.

Filtration

During maintenance, only manufacturer-recommended filters should be used. As the filter ages contaminants are trapped in the filter. After all, that is the job of the filter. Since in most fluid power applications the filter is mounted on the return line just upstream of the reservoir, this inhibits the ability of the fluid to return freely to the reservoir. This restriction to flow creates a differential pressure across a point of restriction. As a result, pressure is held back upstream of the filter, causing a drop in the differential pressure across the actuator and therefore a reduction in flow through the actuator. This will affect both the power of the actuator and the speed of the actuator. Filter design, as well as filter cleanliness, is critical to the performance of the filter, and is therefore critical to the performance of the system. It is important to use only filters that are recommended by the system manufacturer. Less expensive filters often have design shortcuts that cause the filter to change the rate at which it restricts to change as the filter becomes restricted by the particles it is straining from the fluid.

Another significant issue in the choice of the filter is what is called collapse pressure. This is the pressure at which the filtration element itself, not the housing or canister, collapses as a result of the pressure differential (delta P). As the level of plugging in the filter increases, there is a greater and greater pressure differential created across the filtration element. The pressure behind the element increases as the pressure downstream of the element decreases. Eventually, for

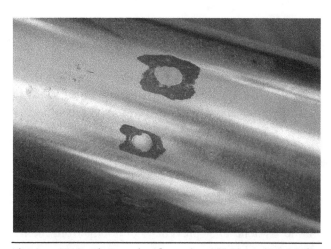

Figure 10-8 The rod of each cylinder should be inspected for evidence of pitting, gouging, and other damage. This type of damage can be indicative of a contaminated system or a system that has been overworked and overheated.

all filters there would be enough pressure differential to cause the filtration element to collapse. This allows unfiltered hydraulic fluid to return back to the reservoir and later be distributed back to the hydraulic system. It also usually results in the filter element itself becoming a contaminating material throughout the hydraulic system **(Figure 10-8)**.

Material compatibility is also important. The filter must be able to withstand the pressures, temperatures, and vibrations to which the system will be subjected. In addition, filters chosen by maintenance people only because of convenience or availability might be made of materials that are incompatible with the hydraulic fluid being used in the system.

During quarterly, semiannual, or annual maintenance, the filters should always be thoroughly checked or replaced. Since in most cases a truck repair facility will not have the ability to thoroughly check or test a filter to see if it needs to be replaced, the smartest thing to do is simply replace the filter. Filters are much cheaper than pumps, motors, cylinders, and valves. Also remember: Just because the filter fits doesn't mean it is the right filter.

Hydraulic Fluids

In many mobile equipment hydraulic systems, the fluid is somehow expected to last through the entire life expectancy of the truck. This is not by design, or planning, but more likely because of a lack of planning. Numerous methods can be used to test the hydraulic fluid for contaminants, additive breakdown, and other damage due to heat and age. This will be discussed later in the chapter.

Many types of hydraulic oils exist. The base stock for these oils can be:

- Castor oil
- Glycol
- Esters
- Ethers
- Mineral oil
- Organophosphate ester
- Propylene glycol
- Silicone

Using oil with the proper base ensures a good life expectancy of the seals, filters, and other soft materials in the fluid power system being serviced. In many jurisdictions, biodegradable hydraulic fluids are required for farm tractors and marine dredging. These hydraulic fluids are based on rapeseed oil, more commonly referred to as canola or vegetable oil. Some hydraulic base systems such as hydraulic brakes on passenger cars and light duty trucks use a hydraulic fluid that is deliberately hygroscopic. This means that when exposed to moisture the brake fluid deliberately absorbs that moisture and in doing so protects the metal components of the brake system from corrosion. Brake fluid is often used in small simple mobile hydraulic systems, both for the convenience of having compatible fluid readily available in the truck repair shop and because of this hygroscopic characteristic. For all these reasons, only the factory-authorized hydraulic fluid should be used when adding to or replacing the fluid of the system. At a minimum, always replace the hydraulic fluid according to the system manufacturer's recommendations. High-temperature environments, dusty environments, and extreme low temperature environments should make the service intervals more frequent **(Figure 10-9)**.

Fluid Analysis

Oil analysis dates back to the early 1940s when the railroads began to use simple spectrographic equipment to monitor the lubricants used in locomotive engines. The purpose was not so much to determine the condition of the lubricant, but rather to determine if contaminants in the lubricant indicated deterioration in major components. In the 1950s, the U.S. Navy began to use the same techniques to monitor the condition of jet engines. Commercial oil analysis laboratories began to appear on the scene in the early 1960s. This made it possible for any service technician or individual equipment owner to monitor the condition of the lubricants and the condition of the equipment being lubricated. Today, many large organizations routinely

Figure 10-9 The cleanliness of the system is often a good indicator of how well the system is maintained. The fluid should be checked for evidence of contamination to be sure it is at a proper level. An inspection should also be conducted to make sure the vent is not plugged or restricted. During these inspections, all safety and warning labels should also be verified since fluid power systems on mobile equipment are often operated by those more familiar with driving the truck than operating hydraulics.

measure the condition of lubricants and contaminants to determine the condition of engines, transmissions, steering systems, and mobile fluid power systems.

As the oil does its job of lubricating and operating, the fluid power system picks up trace metallic particles that result from the normal wear of the machinery. An oil analysis looks at and determines how many contaminants there are in the system and the makeup of those contaminants.

Oil analysis can detect:

- Foreign fluid dilution of the hydraulic oil
- Dirt contamination of the hydraulic oil
- Bearing wear in pumps and motors
- Hard part wear in pumps, valves, and motors
- The misapplication of hydraulic oil

The use of oil analysis on a routine basis can:

- Reduce repair bills
- Reduce catastrophic failures
- Increase machinery life
- Reduce nonscheduled downtime

The substances detected in the hydraulic oil during analysis can help analyze system deterioration in order to reduce unscheduled downtime and turn it into scheduled repair time. In almost all cases, it is cheaper to schedule a repair when equipment is not in use then it is to wait for a failure and make the repair when the machinery or equipment should be making money. Substances that can be detected include:

- Aluminum in the hydraulic fluid can indicate damage to pump or motor thrust washers, pump systems, and bearing surfaces, as well as damage or wear in other components specific to the system being serviced.
- Boron, magnesium, calcium, barium, phosphorus, and zinc are metals normally found in additives of hydraulic fluids. They are constituent components of detergents, disbursements, and extreme-pressure additives. Excessive amounts of these substances in the hydraulic fluid indicate a breakdown in the additive package.
- Chromium does not appear in hydraulic fluid when it is manufactured. If chromium is found during oil sample testing, it is an indication that components found either in the pump or in a hydraulic motor have been damaged.
- Copper and tin are materials used in bearings, bushing pins, and thrust washers. The presence of these two metals may indicate that there have been lubrication problems in the hydraulic pump or in the hydraulic motor which is causing breakdown in these friction-reducing surfaces.
- Iron in the hydraulic oil is indicative that the valves and casings of pumps and hydraulic motors have been damaged.
- Silicon is quite simply dirt or sand. The most likely cause of this is poor filtration of the air pulled into the reservoir, or more likely the system being operated with the reservoir filler cap removed.

In addition to testing for metals, many labs will also do a water-by-crackle test. In this test, the plate is heated to 150°C. A sample of the oil is dropped onto the plate. If the oil contains no water, it will simply smolder on the plate; if the oil contains water, it will crackle and sizzle just like water dropped onto a hot griddle. This test is not quantitative in nature but it does indicate a simple inexpensive procedure to determine if water is present.

Many labs will also test the oil's viscosity. Viscosity, you will remember, is the oil's resistance to flow. Depending on the lab and the type of testing being done, the oil will be heated to 40°C or 100°C and then tested for its ability to flow.

The acidity of the oil is also tested. If high levels of acidity are detected, this indicates either oxidation or contamination of the hydraulic oil.

Other tests such as particle counting, infrared analysis, or Karl Fisher titration may also be done.

AUTOMATED PARTICLE COUNTING

One fluid analysis method is called "automated particle counting." This process is nothing more than what it sounds like. The hydraulic fluid is placed into a chamber where a computer uses a high-resolution video camera to count the number particles in the sample. In times past, this was done by a lab technician using a microscope and a hand counter. The automated equipment makes this process relatively inexpensive and much more practical as a maintenance method. That being said, it is still not inexpensive enough to be done routinely on inexpensive systems.

INFRARED ANALYSIS

Infrared analysis is used to determine the amount of organic compounds in the oil sample. Organic compounds present in the oils will absorb infrared light at specific frequencies. The most common frequencies measured in oil analysis include hydrocarbon compounds that might be picked up from the exhaust system of vehicles, oxidation, water, and glycol.

Modern diesel engines with complex after-treatment systems put out very little soot compared to older engines. Therefore, it is unlikely that much of this soot will be found. Oxidation is a measurement of the degradation of the hydraulic oil's molecules as they combine with oxygen. Oxidation is a normal part of the aging process that can be accelerated as a result of high temperatures and the presence of acids. Oxidized oil does have increased viscosity and can cause damaging deposits throughout the fluid power system. Water and glycol can be present in the hydraulic oil as a result of ruptures in liquid-to-liquid heat exchangers or simply because of careless maintenance.

FILTER TYPES, RATINGS, AND PROCEDURES

A modern fluid power system is subject to contamination from a variety of sources. These sources include built-in contamination that results primarily from the manufacturing process. This contamination usually takes the form of burgers, chips, dirt, dust, fibers, sand, moisture, sealants, welding spatter, paints, rust, and flushing solutions. In fact, any time a new component or replacement component is added to an existing system, that component is a source of new contamination.

Foreign object contamination finds its way into the system through faulty reservoir breather filters, worn cylinder rod seals and wipers, or when new oil is added to the system. In many cases, mobile hydraulic systems are operated in less than ideal conditions. The truck carrying the mobile fluid power system is usually working at a job site that is a dusty, dirty, and even muddy environment. Any time the system is open to oil, the system is exposed to all of these airborne contaminants.

One of the worst sources of contamination is internally generated. As the system is operated, and valves move, pumps pump, cylinders move in and out, and motors rotate, microscopic particles are being separated from their parent material. Most of these particles are as small as 10 to 20 microns in size. A micron is one millionth of a meter or 0.000039 inches. Internal clearances in a high-pressure hydraulic system can be as small as 5 μ (microns). One study showed that poorly maintained systems can have up to 100 million particles per minute flowing through the system.

As a result, good filtration is essential to the life of the fluid power system, and use of the proper filter is critical to good filtration.

FILTER CONSTRUCTION

Most fluid power systems, especially those found on mobile equipment, utilize a filtering strategy known as full flow filtering. This simply means that all of the fluid being removed from the reservoir by the pump will pass through the filter before returning to the reservoir. Some systems will have nothing more than a woven wire screen on either the reservoir's pickup or on the return line to the reservoir. Obviously, a woven wire screen can only filter out the largest particles. When larger particles exist in the system, this is evidence that one or more components have failed. Therefore, the only real function that the screen serves is to prevent large chunks from returning to the pump and damaging it.

Four types of filtration size exist: meshes per linear inch, U.S. sieve size, size of the opening in inches, and size of the opening in microns. **Table 10-1** shows a comparison of the four measurement methods.

As a point of reference, the smallest object observable to the human eye (without the assistance of lenses) is 40 μ.

FILTER MEDIUM

The most common filter medium is plated paper that has been reinforced with a resin or fiberglass to give it sufficient strength to withstand the flow of the hydraulic fluid. This filter medium is then installed on a canister or is part of a disposable metal housing. During the last few decades, the filters contained in the disposable metal housings have given way to a

TABLE 10-1: COMMON FLUID POWER FILTRATION SPECIFICATIONS

Meshes Per Inch	U.S. Sieve Size	Opening in Inches	Opening in Microns
52.36	50	0.0117	297
72.45	70	0.0083	210
101.01	100	0.0059	149
142.86	140	0.0041	105
200.00	200	0.0029	74
270.26	270	0.0021	53
323.00	325	0.0017	44
		0.00039	10
		0.000019	5

replaceable filter that goes inside of a non-disposable plastic or composite housing. This is a result of concerns about the metal housing type filters filling up landfills and becoming an environmental hazard. These paper element filters are normally found on the return side of the hydraulic system just upstream of the reservoir.

Another common type of filter is made up of compacted synthetic fibers. These fibers are not woven but simply compressed into a tight bundle that makes it very difficult for contaminating particulate matter to pass through.

FILTER RATINGS

Other than the composition of the filter medium, the most significant rating for a filter is based on the size of the particle they are designed to remove. Three such ratings exist:

- Absolute rating—This is the smallest particle that the filter will remove from the flowing fluid. This means that a 10-μ filter will not allow any hard particle 10 μ or larger to pass through the filter.
- Nominal rating—This is the size particle that will be removed 98 percent of the time as the hydraulic fluid passes through the filter. Therefore, a filter with a 10-μ rating will allow a particle 10 μ in size to pass through the filter 2 percent of the time.
- Beta rating—The beta rating compares the number of particles of a given size that enters the filter with the number of particles of the same size that leave the filter. Although there is a formula for calculating the beta rating of the filter, the best way to explain it is that a beta rating of 1 means that the same number of particles that left the filter has entered the filter.

TABLE 10-2: COMMON BETA RATIOS AND FILTER EFFICIENCIES

Beta Ratio	Filter Efficiency
1	0%
2	50.00%
5	80.00%
10	90.00%
20	95.00%
75	98.70%
100	99.00%
200	99.50%
1,000	99.90%
5,000	99.98%

The higher the number, the better the beta rating. **Table 10-2** shows the beta ratio as compared to the percentage of efficiency.

LOCATIONS FOR FILTRATION

Screen type strainers are found at the filler neck on the reservoir and on the inlet or pickup that leads to the pump. Not all systems will have these two strainers, if the system does not have a filler neck strain, then the service technician should use extra caution to ensure that dirt and other materials do not fall in the reservoir when checking or filling it. Pressure line filters are located downstream of the pump at a pressure relief valve. If the filter restricts to the point where oil flow is significantly limited, pressure can build up between the filter of the pump. The pressure relief valve is located between the pump and the filter in the event that a severe filter restriction occurs. The most common place to find filters in mobile equipment applications is on the return line just before the returning fluid enters the reservoir.

Performing Routine Maintenance

Performing routine maintenance on a mobile fluid power system is as important as it is on any other part of the vehicle. In many cases, the fluid power system is more important to the value of the vehicle than the vehicle itself. A truck designed to be a snow plow is of little value if the hydraulics that operate the plow unit are inoperative. When new equipment is received, the operator's manual, or a copy of it, should be placed in the vehicle as a reference for the operator. The service manual should be placed on file where it can be accessed by any technician. If the vehicle is to be operated and serviced at a remote location, then a copy of the service manual should be kept at that remote location or in the vehicle. At all times, service and maintenance suggestions made by the manufacturer should be followed precisely. Service intervals suggested by the manufacturer should also be followed precisely. What follows is a general overview of what the maintenance guide will suggest in the operator's manual and/or the service manual.

THE FLUID

Inspect the fluid for signs of aeration and hard-particle contamination. If the system is operated in an environment where there are other fluids and gases present, a small sample should be taken and sent to an oil analysis lab. Signs of bubbles or a creamy appearance to the oil would be signs of aeration. Take a small sample of the oil from the reservoir and run it between the thumb and forefinger to feel for grit. Keep in mind that even the smallest amount of detectable grit can indicate severe contamination. This is because even though human fingers are highly sensitive, they do not have the touch resolution to recognize a particle just a few microns in diameter. And yet such particles, large enough in number, can significantly damage fluid power systems. Therefore, if particles can be felt, it is time to flush the system and replace the fluid.

Flushing the system does not require the use of any special fluids, cleaners, or detergents. Simply drain the system and refill according to the manufacturer's recommendations. Afterward, drain the system and refill it according to the manufacturer's recommendations again.

THE HOSES AND LINES

Next, allow the system to warm up. Operate the system throughout its range of operation several times

and inspect the hoses and lines for twists, kinking, dents, cuts, and abrasions. Next is a heat gun, not your hand, to inspect for changes in temperature.

Note: Use of the hand to check for changes in temperature while the system is under pressure is very dangerous. Fluid escaping from hoses under high pressure can often be invisible. These high-pressure leaks can have sufficient force and velocity to enable the hydraulic fluid to penetrate the skin. This can not only cause serious cuts but can also lead to toxic responses and serious infections.

STICKING AND BINDING HYDRAULIC PARTS

Effective operation of fluid power equipment requires precision control. If control levers, or the valves they control, are sticking or binding, the operator will have difficulty working the machine in a safe and productive manner. Further, damage to actuators can result in operation that is not smooth and is therefore unsafe. The equipment should be operated by the technician throughout its range of operation several times while watching for smoothness of operation. Keep in mind that the hydraulic components themselves can stick, but it is equally likely, if not more likely, that the arms, pulleys, and other devices operated by the hydraulic cylinders and motors could be sticking as well. Knuckles, pivot pins, and other articulation points and lube points should be inspected for wear, damage, and lack of lubrication.

THE FILTER

When it comes to maintenance decisions about servicing, filters are easy. The manufacturer's recommendations should be followed. There is no doubt that a filter is much cheaper than the hard-metal components and seals that could be damaged due to poor filtration.

LEAKS

Hydraulic oil leaks are not only messy and have a potential for internal impact, they can also prevent proper operation of the system. There are basically two types of leaks: external leaks and internal leaks. External leaks may or may not be readily visible. Rarely would there be a stream of hydraulic fluid projecting from a defective hose, fitting, or component. In some cases, the leak is very slow and the only evidence will be dirt and debris that has adhered to the moisture of the oil surrounding the lake. Other times, especially under pressure, the oil may be vaporizing as it leaves the hose, rendering the position of the leak all but invisible (Figure 10-10).

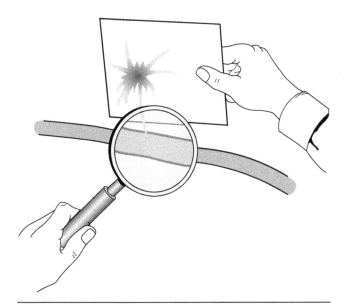

Figure 10-10 Checking for leaks may seem like a simple and relatively safe task. However, if any part of the system retains pressure while the test is being done, high-pressure hydraulic fluid could be jetting out from any leak that did exist. Very small leaks may not be visible to the naked eye, yet if the technician runs a hand across the jet, it could cut into the flesh just like a knife. A far safer method is to run a piece of paper across the suspected lines as shown here. Any jet of high-pressure oil will damage and stain the paper instead of the technician.

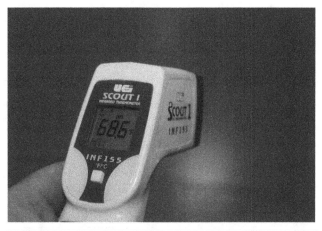

Figure 10-11 Since hot spots in fluid power systems can be indicative of restrictions, worn components, or other types of damage, it is always a good idea to inspect the hoses and components for excessive temperature. At one time, this was done through the rather dangerous method of touching the components. Today, it is far safer and more accurate to use a device such as the infrared heat gun shown in this photograph. The better ones have a laser that is able to pinpoint the spot where the temperature is being tested when the trigger is pulled. These tend to be a little more expensive than the one without the laser, but are well worth the extra investment.

Continue the inspection of the system by looking for leaks in the lines and hoses. Once again, use of the hand or another body part to check for leaks is extremely dangerous. One very effective method of looking for leaks is with the use of a piece of heavy white paper. Run the paper all along the lines, hoses, and other components, making sure to keep hands, arms, and other body parts well away from any of the components. When a leak is present, it will immediately stain the piece of white paper. Remember that it is important to be operating the equipment while testing for these leaks. Many leaks will only present themselves when the machinery is operating under a load. Therefore, if the leak does not readily present itself, the technician should ask a co-worker to operate the equipment under about a 90 percent load while they check for leaks.

Internal leaks are trickier to find. Evidence of an internal leak will be a motor or a cylinder that will not hold its position when there is a load and it is parked or there is slow operation. The most likely places for internal leaks are motors, cylinders, pumps, and any of the valves. Use a heat gun to look for abnormal operating temperatures within specific components,

which would indicate a leak exists **(Figure 10-11)**. More on this topic can be found in the troubleshooting chapter **(Chapter 11)**.

OVERHEATING

A final, yet critical part of a routine inspection should be to look for anomalous temperature changes. These temperature changes can be evidence of internal leaks, excessive air or moisture in the system, excessive pressure, or a failure of any heat exchanger to do its job properly. Do not forget that the primary heat exchanger in most mobile fluid power systems is the reservoir. The most common reason that the reservoir is unable to do its job in exchanging is that there is insufficient fluid in the system. Another common cause of system overheating is servicing the system with improper oil or fluid. In many cases, even simply topping off the system with the wrong fluid can cause the system to overheat.

It has been found that if there are no unusual temperatures, the condition of the heat exchanger should be checked. Be sure there is nothing to interfere with airflow, including a damaged core or fins **(Figure 10-12)**.

Inspect the fluid	Proper fluid level?	Yes	No
	Evidence of aeration?	Yes	No
	Evidence of hard particle contamination?	Yes	No
Inspect the lines and hoses	Evidence of twists, kinks, dents, or cuts?	Yes	No
	Evidence of overheating?	Yes	No
Inspect controls/components for sticking and binding	Evidence of sticking controls or control valves?	Yes	No
	Evidence of sticking actuators or actuated devices?	Yes	No
Replace the hydraulic fluid filter	Is the recommended service interval approaching?	Yes	No
Inspect the system for leaks	Are there any internal leaks?	Yes	No
	Are there any external leaks?	Yes	No
Inspect the system for overheating	Insect the heat exchangers for air flow and dirt	Yes	No
	Is the operational temperature of the system acceptable?	Yes	No

Figure 10-12 Checklist for a basic visual inspection.

Review Questions

1. Routine maintenance is not required on some fluid power systems because their service cycle is such that they are only used 20 or 30 minutes per day.

    A. True   B. False

2. Simple fluid power equipment such as this engine hoist only needs to be inspected:

    A. Annually.   C. Weekly.

    B. Monthly.   D. Daily.

3. Heat exchangers need to be inspected for all but which of the following?

    A. The proper pressure   C. Leaks

    B. The proper airflow   D. Fan operation

4. Technician A says that the filter should always be replaced according to the manufacturer's schedule. Technician B says that the manufacturer's recommendations show this is the minimum that should be done, and under extreme operating conditions maintenance should be done more frequently. Who is correct?

    A. Technician A only   C. Both Technician A and Technician B

    B. Technician B only   D. Neither Technician A nor Technician B

5. Electrical wiring harnesses and connections should be inspected routinely. Among the things the technician should look for is evidence of corrosion. Corrosion should be cleaned immediately when found, because if left long enough corrosion can cause:

    A. The wires to break.   C. Damage to the wires.

    B. High resistance in the wires.   D. All of the above

6. A cracked or damage safety switch should be replaced immediately because:

   A. It can keep the system from operating.

   B. It can keep the system from shutting down when the system is operating outside of its design safety parameters.

   C. It can leak hydraulic fluid.

   D. It is not necessary to replace safety switches because the hydraulic system is capable of preventing overpressurization and other problems that can cause damage or catastrophic failures to the various components.

   E. All of the above

   F. A and B

   G. A, B, and C

7. One of the advantages of an electronically control system is the ability to self-diagnose and tell the system operator, as well as the technician, exactly what type and level of service is required.

   A. True

   B. False

8. Oily dirt around line settings and other components can be indicative of:

   A. Poor maintenance.

   B. Hydraulic oil leaks.

   C. Old and seeping hoses.

   D. All of the above

9. The proper fluid level in a hydraulic system reservoir is essential because of which of the following?

   A. A low fluid level can cause overheating in the system because there is not sufficient rest time in the reservoir for the oil to give up enough heat.

   B. A low fluid level can cause the pump to draw air, and therefore cause erratic operation of the system.

   C. A low fluid level can concentrate the sediment and other contaminants in the oil, making them more hazardous to the system.

   D. With adequate filtration and service of the filter, contamination due to a low fluid level will do no harm.

10. An infrared gun is used while performing maintenance and inspection of the system to determine:

    A. The pressure in the system.

    B. Whether or not the reservoir is at the proper level.

    C. If there are hot spots anywhere in the system that could be caused by restrictions.

    D. An infrared gun serves no purpose with a fluid power system.

CHAPTER

11 Troubleshooting Fluid Power Systems

Learning Objectives

Upon completion and review of this chapter, the student should be able to:

- Describe the troubleshooting process.
- Interview the system operator to gather details for troubleshooting purposes.
- Design a troubleshooting plan.
- Determine possible causes of symptoms such as high temperature, shuddering and vibration, and other common symptoms.

Cautions for This Chapter

- Installation of flow meters and pressure gauges should only be done when the system is powered down and pressure has been relieved from the component.

- Always wear the proper personal protection equipment (PPE) when troubleshooting a fluid power system.

- Always assume that the component being worked with and tested may fail at any moment, resulting in a high-pressure discharge of hydraulic fluid.

- A malfunctioning fluid power system can experience high temperatures at various points and in various components throughout the system. The temperature of these components should not be checked by touch but rather with an infrared heat gun.

- A malfunctioning fluid power system may have weak hoses and lines in the system. The technician working on this with a functioning system should always wear protective eyewear, clothing, and gloves to reduce the chances of injury.

- A malfunctioning fluid power system can suffer from catastrophic leaks at any time. While the system is operating, and especially while testing the system, the technician should protect himself or herself from direct exposure to lines and components that may suddenly rupture.

- Hydraulic fluid under pressure can be injected through the skin. The technician working on a malfunctioning system should never place his or her hands or other body parts on or near any components while testing the system.

- Before affecting any repairs, the service technician should follow proper lock out/tag out protocols as outlined by the safety officers within his or her organization.

- Spilled oil can be very slick and hazardous to walk on or handle equipment on.

Key Terms

contamination	flow meter	pressure gauge
flow meter pressure tester	interview	

INTRODUCTION

Troubleshooting fluid power systems is far simpler than troubleshooting many systems on today's motor vehicles. But like any troubleshooting, good solid deductive procedures and reasoning are required to make an accurate diagnosis. A logical, and consistent, step-by-step approach is essential in making the right diagnosis the first time. All technicians are tempted to go by gut feel or to utilize previous experience in making their troubleshooting decisions. Both of these are an important characteristic of a good fluid power diagnostician. However, they should be used as a valid starting point and testing should be done to verify that that gut feel or previous experience is correct.

One of the biggest timewasters in troubleshooting is running around looking for information related to the system the technician is about to diagnose. In a well-organized shop, there is a room set aside as a library that contains all of the data books, troubleshooting guides, and owner's manuals for the equipment the shop maintains. When any mobile fluid power system comes in for service, diagnosis, or repair, the technician knows where to go to find all the information available on that system.

Today, many manufacturers' fluid power systems have online reference materials that can be accessed via the Internet by authorized service technicians. These online materials are usually in a format similar to Adobe PDF files that allow the technician to open, read, and utilize the materials by means of a free software reader. This also allows the manufacturer to make updates and corrections on a routine basis. When working on a piece of equipment that is manufactured by a company that has an online service literature site, the technician should take full advantage by checking for updates to specifications and diagnostic procedures, and checking for corrections.

To be blunt, the biggest enemy of successful troubleshooting is arrogance. As service technicians, we are constantly bombarded by engineering issues that often result in repeated patterns of failure. When we begin to think that a specific set of symptoms always means that the same component or subsystem is at fault, we will find ourselves trying to prove what is defective rather than trying to eliminate that which is working properly. One can waste a lot of time trying to prove something is defective when it is not.

TALK TO THE OPERATOR

The first step of any successful troubleshooting procedure is to gather as much information as possible about the symptoms. A wise old technician once said, "If you talk to the operator long enough and ask the right questions, the operator will tell you exactly what is wrong with the equipment." If enough of the right questions are asked, and it is a technician who understands how the system operates, then the operator will have helped the technician eliminate 80 or 90 percent of the things that could not cause the symptom. With only 10 percent or so of the possible causes remaining, the technician is able to target or pinpoint the exact components or subsystems to look at first.

In many shops, the technician never has the opportunity to speak to the operator. Often, the shift worked by the operator and the shift worked by the technician do not coincide. If this is the case, then the technician must become the operator. Hopefully, whoever wrote the repair order has written down as concisely and as accurately as possible the operator's complaints and concerns. With this information, the technician must become the equipment operator. The vehicle should be moved to where the equipment can be safely operated and then the technician should operate the equipment in such a way as to maximize the opportunity for the symptoms to occur.

A good troubleshooting **interview** will have five basic questions answered:

- Who?
- What?
- Where?
- When?
- How?

"Who" may seem like a question with an obvious answer, but keep in mind that the person delivering the vehicle or equipment for repair may not be the same person who experienced the problem. Also, to be honest, equipment operators have various levels of skill. In a fleet shop, one gets to know the equipment operators and their habits quite well. Also, in most independent service provider shops, there is a relatively consistent customer base. This also allows the technician to get to know the equipment operators quite well. In some cases, the operator has an unrealistic expectation of what the fluid power system is able to do. Therefore, the symptom or complaint that has the operator concerned may be nothing more than operator error. This does not mean that the problem should not be investigated. The old legend about the boy who cried wolf sometimes becomes reality. Remember in the legend that the boy hollered that he was being attacked by wolves and the villagers came to his assistance. He did this so many times that when he actually *was* attacked by wolves none of the villagers believed him. Equipment operators may complain about equipment problems over and over when there actually is no problem, and then when they complain about a real issue, perhaps even a safety issue, the technician may ignore the operator's concerns. No matter how consistently the equipment problem has turned out to be due to operator error, or the operator expects too much from the equipment, each complaint concerned must be thoroughly checked.

"What" is a precise description of the symptom or concerns experienced by the operator. The difficulty in this part of the troubleshooting procedure is that often equipment operators and equipment technicians do not speak the same language or use the same terms. In fact, when the same terms are used, the definitions of those terms may be different. If possible, ask the equipment operator to demonstrate the problem the equipment is having. If the operator is not available, the technician should operate the equipment in as close to the same environment and manner that the operator had done. Once the precise concern is determined and simulated, the technician has a solid set of symptoms from which to troubleshoot. Basically, the technician should simulate the problem.

"Where" is often as important as "What." The environment in which equipment is operated has a major impact on how the equipment performs. Ambient temperature, humidity, and the amount of airborne particulates, such as sand and dust, can significantly affect how the equipment performs. For electronically controlled fluid power systems, the proximity to radio signals and intense electromagnetic fields can also significantly alter how the equipment performs. For instance, if the equipment is being operated near airport radar, communication equipment can affect the operation of the microprocessors in the electronic control systems.

"When" is an indication of how the heat generated through the operation of equipment affects the operation of the system. It can also give a clue as to the condition of the hydraulic fluid. If the system operates normally when cold and continues to operate normally for some time as the system reaches normal operating temperature but then the operation deteriorates as the system is stressed, it is likely that heat is affecting the operation of the system.

If the malfunction begins as soon as the machinery starts operation and then begins to work properly once it approaches operating temperature, this could be an indication of degraded or oxidized oil. In the case of electronically controlled systems, it could be a fault in the programming of the controller.

If the malfunction appears and disappears, then it is an intermittent problem that is not temperature-dependent and the technician should attempt to troubleshoot it while the malfunction is present.

If the malfunction begins as soon as the machinery starts operation and continues throughout all cycles of operation, this is good news for the technician. One of the most difficult aspects of troubleshooting is the fact that the fault often does not present itself while the technician is attempting to operate the equipment. A malfunction that occurs continually makes the troubleshooting easy.

"How" is the most difficult information to get accurately from the operator. In many cases, the fault presents itself when the equipment is being abused or misused. It is unlikely that the operator will be entirely forthcoming with information about how he or she is operating the equipment; this is true even when they are operating the equipment properly, and even more true when they are not using the equipment properly. Basically the "how" part of the question relates to how the equipment is being operated. One of the most difficult aspects of equipment operation for equipment operators to grasp is the concept of duty cycle. All hydraulic equipment has a duty cycle. This is how long the equipment operator can use the equipment versus how long the equipment has to rest. This is often expressed in a duty cycle. If the equipment has a duty cycle of 10 percent, then the equipment can be used for six minutes and rested for sixty minutes. The operator's manual will inform the technician and the operator how long the equipment can be operated uninterrupted before it must be shut off and allowed to

Troubleshooting Interview

- Who was the operator reporting the problem?
- What was the problem with the equipment?
- Where was the equipment when it happened?
- When did the problem occur?
- How was the equipment being operated?

Figure 11-1 The best source of troubleshooting information is the person that first experienced the equipment failure. There are several critical questions that should be asked.

rest. The preceding example shows a continuous use time of six minutes. This is only an example since continuous use can vary from seconds to hours or even days. If the continuous use factor is one hour and the duty cycle is 50 percent, then the equipment can be operated for an hour and then rested for an hour. Exceeding the continuous use rating for the duty cycle of the equipment can cause premature wear and expensive failures **(Figure 11-1)**.

Gather Information

Much of troubleshooting does come down to analyzing the information received from the equipment operator and the data received from testing. In order to make this analysis, it is essential that the technician understand how the system is supposed to work and what components are used to make the system operate. When these two aspects of understanding are brought together, then the technician was able to determine what component or components have effects that are keeping the system working properly. So the goal of gathering information is to gain an understanding of how the components work together when the system works properly.

System manufacturers provide a variety of information for helping the technician. The most important of these is the system diagram. **Chapter 4** and **Chapter 9** of this book were included to assist the technician in gaining this understanding. System manufacturers may also supply a diagnostic sequence or troubleshooting book. These will often have diagnostic flowcharts that begin with the symptom, or in the case of electronically controlled systems, may begin with a diagnostic code. These flowcharts are designed to assist the technician in making a logical sequence of tests and analyzing the results of those tests in a step-by-step process. The operator's manual should also be gathered to ensure that the "problem" with the equipment is not simply due to an operational quirk designed into the system. For instance, there may be a non-intuitive safety valve that has to be moved or

activated in order for the equipment to perform a specific function.

Once the information has been gathered, the technician should find a quiet place where the information can be spread out—for instance, a table in the break room. The technician should get a cup of coffee and then sit down and study the information until they feel confident there is a thorough understanding of the components used and their interaction. In this scenario, the table in the break room symbolically suggests a quiet place where the technician can sort through the information they have gathered. The cup of coffee symbolizes ignoring distractions and taking the time to gain the needed understanding. This should be done before the technician applies the first diagnostic tool to the system or even opens their toolbox.

Perform a Visual Inspection

Never overlook the basics. There are times when the experienced technician can get on a diagnostic track based on a set of assumptions resulting from testing they have already done. Never be embarrassed or shy about asking a co-worker to either rethink your thought processes or begin the diagnostics from scratch. Also keep in mind that the top technicians in the shop each have a common way of analyzing a problem or defect. A new viewpoint, even one based on limited experience, can often help the technician find the source of problems.

The following bullet points are suggestions on what should be done at a minimum during a visual inspection **(Figure 11-2)**.

- Check the hydraulic oil level and type.
  - Is the level too high or low?
  - Is the correct oil being used? Be sure to check this against the manufacturer's specifications.
- Check the hydraulic oil condition. Is it foamy, milky, burnt, or dirty?
- Check for hydraulic oil leaks.
- Look for signs of overheating. Inspect for evidence of burnt paint or discolored metal on all of the hydraulic components, not just those where you would typically expect to find pressure.
- Check for damage such as dented cylinder barrels, bent or scored cylinder rods, crushed hydraulic tubing, collapsed hoses, and so on.
- Inspect to see that the system has been properly maintained.
  - Look at the grease points for evidence of proper lubrication.

Figure 11-2 One of the most essential elements in troubleshooting a fluid power system is the visual inspection. Many defects leave telltale evidence of their existence. This can be in the form of shiny areas in articulating components, black stains that indicate a mixture of environmental dirt and oil, or leaks. Additionally the use of the information gathered during the operator interview can help the technician and point the area to begin the complex troubleshooting but also the visual inspection.

- Check service records to confirm that the filters have been replaced at proper intervals.
- Check linkage adjustments.
- Have there been machine modifications?
  - Has the bucket or box size been increased?
  - Has equipment been retasked to perform a job for which it was not designed?

Although it is not always necessary, it is nevertheless a good idea in many cases to remove the hydraulic oil filter that has been in service and inspect it for metallic **contamination**. This contamination would be strong evidence that a component has failed.

Perform Operational Testing

Operational testing is a very important part of troubleshooting. This is especially true when it is not possible to interview the operator. During operational testing, the technician becomes the operator.

Before beginning an operational test, be sure that the hydraulic oil has been brought up to operating temperature. The only exception to this would be when the equipment has an operational fault only when cold, and the equipment is of a type that must be pressed into service regardless of the temperature, such as rescue or emergency equipment.

Once the equipment is up to operating temperature, the technician can begin an operational test of the equipment. In a typical system, the operating temperature is between 150°F (65.50°C) and 180°F (82.20°C). For any given piece of equipment, the operating temperature can vary greatly; therefore, during the information-gathering phase of the diagnosis, a technician should find out from the manufacturer's data what the normal operating temperature is. Once the system oil is at normal operating temperature, it is time to put the machine through its paces. Again, exactly what to do to put the machine through its paces should have been researched during the information-gathering phase. This information can be found either directly in the form of daily operator testing or implied in the operational parameters outlined in the operator's manual.

If any of the machine functions do not perform, the technician should troubleshoot a portion of the circuit that is misbehaving. This is where the system schematics become invaluable. Essentially three circuit defects can prevent a function of the machine from happening.

- Oil ceases to flow through the actuator: blockage.
- Oil is diverted away from the actuator: internal or external leak.
- There is insufficient pressure or flow to allow the function to occur.

At this point, the technician should sit down with a schematic and determine how the oil must flow through the circuit to make the function happen. The following are some important steps:

1. Starting at the pump, the technician should trace the flow of the hydraulic fluid and isolate the branch in the system where those things that are working branch off from those that are not. Everything up to the point where the good branches off from the bad can be assumed to be working properly. The technician can now concentrate on those components in the branch that contain the actuator or component that is not operating.
2. The technician must locate the actuator controls on the diagram. These controls are usually in the form of manually operated valves, manually operated directional control valves, and pneumatically or electrically operated control valves. The technician should verify that these controls move and operate correctly.

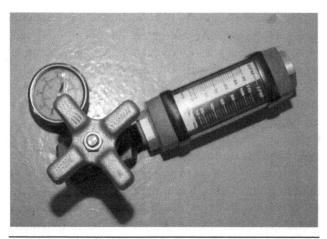

Figure 11-3 In much the same way that the digital volt-ohm meter is the basic tool for troubleshooting an electrical system, the flow meter pressure tester is the basic tool for use with a fluid power system. The tool consists of two gauges, one to measure pressure and the other to measure flow rate. It is usually connected in a line running from the pump and heading toward the actuator. Exactly where it is hooked up depends on what the technician is attempting to discover. The typical connection also places the manual shutoff valve of the tester downstream of the pressure gauge. This will allow the technician to shut off flow, yet be able to measure the pressure inbound to the tester from the pump. When the manually operated valve is closed to flow, the pump is dead-headed and the maximum pump output pressure can be tested.

3. The technician should install a flow meter pressure gauge between the control valves and the actuator **(Figure 11-3)**. The technician should then observe the flow meter pressure gauge as the control valves are operated and observe changes in pressure and flow rate. If the pressure between the control valves and the actuator remains quite low, and well as the flow rate, then there is a restriction to the hydraulic fluid flow between the common point where the good splits off from the bad and the point where the flow meter pressure gauge is installed. If, however, the flow rate is high and the pressure is low, a leak likely exists between the common point where the good splits off from the bad and the actuator. The condition of the actuator can be confirmed by moving the flow meter pressure tester to the return side of the actuator. If there is a relatively normal flow on the return side of the actuator, this is an indication that the actuator is defective and bypassing internally. If there is no flow or low flow on the return side of the

actuator, and there was good flow to the inbound side of the actuator and would indicate that fluid is bypassing the actuator. The schematic and the system itself should be studied to locate the alternative paths through which the fluid must be flowing.

This is simply an example of how the hydraulic schematics **(Figure 11-4)** can be used to pinpoint circuits that need to be diagnosed based on how the system is performing.

Actuator Cycle Times

Actuator cycle times are measurements of how long it takes for an actuator such as a piston to go through its entire range of motion, both extending and retracting. Keep in mind that those system specifications that were found during the information gathering phase are essential to knowing whether the actuator is performing properly. Remember also that the normal extend and retract times can be different. The same cycle time testing applies if the actuator is a motor. The actuator time specifications are usually found either in the operator's manual or in a troubleshooting guide. Slow cycle times are always an indication of insufficient flow. To ensure that the test is done accurately, the actuator should be cycled several times and the cycle time measured with a stopwatch.

Check Hydraulic Stall Speed

Hydraulic stall speed is a measurement of how fast the engine is turning when a hydraulic system is at maximum load. The speed is checked with a tachometer when a hydraulic actuator is stalled, when a hydraulic cylinder or motor is attempting to lift more than it is capable of, and the throttle of the engine is in the wide open position. The amount of RPM drop is proportional to the amount of power drawn from the engine. If the RPM drops below the rated level found in the manufacturer documentation then:

- The engine lacks sufficient power to handle the load, air filters, and exhaust restriction, and so the engine condition should be checked.
- The pressure relief valve is just higher than the specification, allowing the fluid power system to put an excessive load on the engine.

If the engine speed does not drop to a level close to the manufacturer's specification, the problem could be:

- The engine power has been increased either through upgrading of components or, in the case

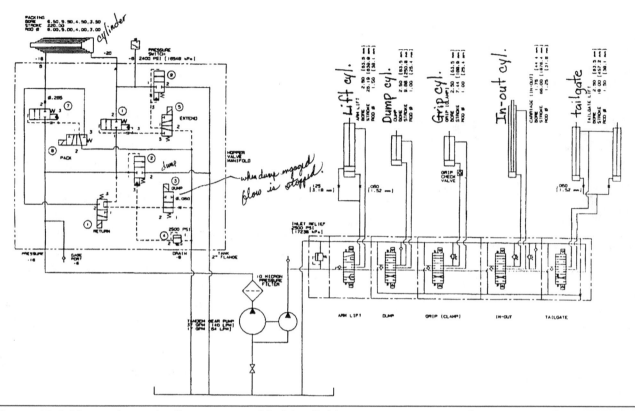

Figure 11-4 Many schematic diagrams provided by fluid power system manufacturers are photocopies of photocopies printed in a book or pamphlet. The images are often blurry, unclear, and even appear to be smudged. In this example, the technician has attempted to compensate for the entire nature of the diagram by making notes. Before sitting down to study the diagram, the technician should, if possible, make a photocopy on which the technician can trace hydraulic flow lines, make notes, and use a highlighter. The more complex the system, such as this trash compactor, the more important it is to do this.

of many diesel engines, programming changes in the engine control module.

- The relief valve has been adjusted below the specification, creating a reduced load on the engine.
- The pump is damaged and unable to produce sufficient pressure.
- The motor or cylinder is damaged and is bypassing internally.
- There is an internal or external leak preventing pressure from building to the specified level.

The preceding list is not a complete list of all of the conditions that can cause the hydraulic stall speed to be incorrect. Operational tests, actuator cycle times, and stall tests can be done quickly and without looking up very many test tools. The next step would be to test with hydraulic test equipment using a **flow meter pressure tester**. With this tester, the following tests can be done on the system:

- Pressure testing
- Neutral pressure
- Working pressure
- Flow testing the actuators for internal leakage
- Flow meter testing

Pressure Testing

Using a flow meter pressure tester can be a slow and tedious process. Often the gauge will have to be connected and disconnected in several positions throughout the system. Therefore, proper planning and analysis of the schematic diagram can minimize the number of times the gauge will have to be connected and disconnected **(Figure 11-5)**. It is important to select the proper gauge and connections when testing a system. Choose a **pressure gauge** that has a scale with a range that is approximately twice the maximum system pressure specified in the manufacturer's data. This allows the gauge to be operated in its midrange where it is most accurate and also prevents damage to the gauge. Be sure that all the fittings, hoses, and connections are capable of withstanding twice the pressure you expect to measure.

Figure 11-5 Pressure testing can be used to test the condition and operation of not only the pump but also of the pressure regulating and limiting components in the system. When pressure is tested at two different points, this can be used to isolate points of restriction within a hydraulic system.

If only one component fails while pressure testing, serious injury could result.

Most manufacturers of larger and more complex mobile hydraulic systems provide pressure taps in the system to make installation of a pressure gauge easier. During the service information search, the technician should make an effort to locate these taps and utilize them whenever possible. Before installing any pressure gauge for a flow meter, it is essential that the technician clean the area around the tap thoroughly before opening up the system, and make sure that all of the fittings and threads are in good condition.

Note: Many times an experienced technician has taken the lazy route and put him and others at risk by using substandard fittings to make a pressure test connection.

CAUTION *It is critically important before installing a gauge to make sure the hydraulic system is depressurized. Hot hydraulic oil under pressure can cause serious personal injury.*

One step that should never be skipped in an open center hydraulic system is checking the return flow side of the system. From this connection point there are three critical pressure measurements that can be made in an open center hydraulic system:

- Neutral pressure
- Working pressure
- Maximum system pressure

For the initial pressure check the gauge is connected parallel with the pump outlet line. In this position, the gauge reading indicates if there is any resistance to flow downstream of the gauge. In this case, that would be through the open center of the directional control valve, through the return line, through the filter, and back to the reservoir. Essentially, there should be no pressure. If there is, then the condition of the return line filter and the return line hoses should be checked. If the hoses and filter are in good condition, the reservoir should be tested to see if it is holding a pressure. Unwanted pressure in the reservoir can cause pressure to build up in the return line.

Working pressure refers to the pressure required of the actuator with no load. If you are working on a refuse packer truck, for instance, extend the packing cylinder with nothing in the bed. Then, retract the packing cylinder with nothing and the pressure readings will show:

- The resistance to flow in the lines and fittings
- The pressure required to send a packing cylinder
- The effect of any friction on the movement of the packer

Since the amount of force required to extend the packing cylinder is constant, any excessive working pressure can indicate restrictions in the lines and fittings, or excessive friction either inside the actuator or on the blade of the packer.

Maximum system pressure occurs when a maximum load is placed on the actuator **(Figure 11-6)**. The manufacture of the equipment would probably have guidelines on how to measure maximum system pressure for the equipment with which you are working. In the example of the refuse packer truck, either the packing cylinder will go to full extension and deadhead itself or the manufacturer will provide a tool to limit the extension of the packing cylinder or provide recommendations on what to place in the bed or hopper of the truck to force the packing cylinder and blade against to limit extension. Maximum system pressure is measured by stalling the actuator while the engine is running at full throttle and taking a pressure reading.

- If the maximum system pressure is too high:
  - There is a faulty relief valve.
  - There is a misadjusted relief valve.
  - There is a restriction in the outlet from the relief valve.

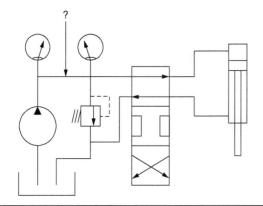

Figure 11-6 A difference in pressure between two points when fluid is flowing indicates a restriction. In most cases, the restriction is supposed to be there either as a result of the device being specifically designed to cause a pressure reduction or limit the flow, or as a normal restriction as fluid flows through the house and fittings. In this example, the diagram indicates only a hose or other piping between the two gauges; therefore, there should not be any restriction. The fact that there is a difference in the readings of the two gauges indicates there is a restriction to flow located between the two gauges. An infrared heat gun can be used to verify there is a restriction and to pinpoint the exact location of the restriction. The hose would tend to be hot upstream of the restriction and noticeably cooler downstream of the restriction.

- If the maximum system pressure is too low:
  - There is a faulty relief valve.
  - There is a misadjusted relief valve.
  - The pump is bypassing internally.
  - The actuator is bypassing internally.

Flow Testing

Flow testing is done when the operational tests indicate the actuators are operating slow. One of the simplest ways of performing a flow test is by checking for components that are heating rapidly. Rapid increases in temperature during the operation of the system indicate that the component is bypassing hydraulic fluid internally. Many people use their hands to check for this rise in temperature. While this is an effective method, it can also be dangerous. A far better way to do it with today's technology by using an infrared heat gun. Many of these have a laser sight on them that makes it easy to pinpoint the area where the temperature is being checked. It is also possible to check components from a distance, which means that many components can be

checked in a relatively short period of time. If the component is detected to be heating rapidly, then an internal leakage test should be performed.

- Testing cylinders:
  - Fully extend the cylinder so the piston has reached its maximum point of travel.
  - Remove the line from the return side of the cylinder. Be very careful to ensure it is the return line and not the pressure line.
  - Place the line in a container.
  - Move the directional control valve to the extend position with the engine at full throttle.
  - Measure the amount of oil that flows out of the line for 10 seconds.

Any oil that flows out of the return line and into the container is oil that is leaking past the piston seals or packing. The manufacturer may or may not have a specification for allowable leakage past the seal. Therefore, a little common sense may need to be used. For instance, if a quarter of a gallon was gathered in the container in 10 seconds, then six times that much would have been gathered in a minute. Without a doubt, a gallon and a half would be excessive bypass on almost any system, certainly on any mobile hydraulic system.

The same process can be followed for testing bypass in a motor. An amount of weight would have to be placed on the motor so that it stalls when the engine is at full throttle. The amount of oil flowing from the drain line or return line of the motor at this point is the amount of leakage in the motor. Again, the manufacturer may or may not have an allowable specification. However, it can be safely said that only a minimal amount of fluid should be gathered in the container from the drain line or return line of the motor.

Remember that the technician was led to this test procedure by slow actuation during the actuator cycle time test. Since minor bypass leaks will not show up during the cycle time test, the amount of bypass fluid that would be gathered in the container would be significant.

Another indication of an internal bypass or actuator problem could be the failure of that actuator to hold its position. For instance, if a cylinder or motor is stopped at a given position with a significant weight on it, and the weight slowly begins to pull the actuator—for instance, a cylinder—to its retracted position, this could also indicate internal bypassing. This symptom is referred to as "drifting."

Flow Meter Testing

In more complex systems, a **flow meter** is the central troubleshooting tool **(Figure 11-7)**. However, before the technician races off to get the flow meter to start the testing procedure, a few points should be considered **(Figure 11-8)**:

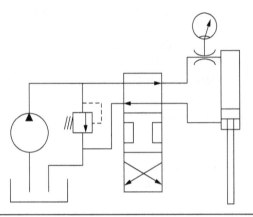

Figure 11-7 In addition to testing for flow rate while a component such as a cylinder or piston is moving, a flow meter can also be used to test for flow when flow should not be occurring. In this example, the cylinder extend side has been filled with fluid and the piston has moved to its fully extended range of movement. At this point, there should be no flow to the cylinder. In fact, a pressure relief valve should have opened and the fluid should simply be flowing through the pressure relief valve back to the reservoir. The fact that there is a flow indicated with the piston fully extended shows that there is fluid passing from the extend side into the retract side of the cylinder. This is probably caused by bad piston seals or a damaged piston.

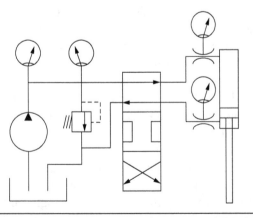

Figure 11-8 Because whatever flows into the cylinder must also flow out of the cylinder, the flow meter can be placed on either the extend side or the retract side to test for internal bypassing in the cylinder.

- Installing and using a flow meter is much more time-consuming than performing operational tests.
- Connecting a flow meter into a system requires that the system be opened up, which means it then runs a risk of contamination.
- If the flow meter was last used in a system that experienced a significant failure, the meter itself may be contaminated.

Analyze the Test Results and Draw a Conclusion

Compared to many systems on modern trucks, troubleshooting a hydraulic system is relatively simple. In fact, the bulk of the troubleshooting process, and the most important information obtained during it, is done by simply familiarizing oneself with the system during the information-gathering phase. Taking the time to understand how the system is supposed to operate can vastly decrease the amount of time it takes to troubleshoot and draw a conclusion **(Figure 11-9)**. When analyzing the information, the following should be considered **(Figure 11-10)**:

- Does the information that has been gathered indicate a flow problem?
  - Is the actuator operating slowly?
  - Is the actuator drifting?

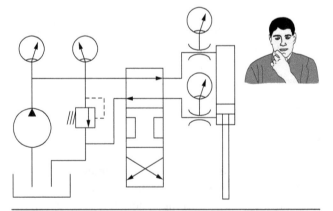

Figure 11-9 Probably the most critical stage in troubleshooting a fluid power system is the stage where tests have been performed and the technician sits down, studies the test results, analyzes them, and makes a diagnosis. This step is critically important because often what appears to be the source of the problem at first glance is actually a symptom of the problem, not its root cause. The best technicians always think in terms of root causes. Once a root cause has been determined and repaired, then it may be necessary to do further repairs or maintenance on components affected by the problem.

Figure 11-10 Seeing a box of components like the one shown in this fluid power specialty shop is pretty normal. In fact, there were several boxes of this type in the shop where this photograph was taken. However, a box of miscellaneous components such as this at a shop that does a wide range of work on mobile equipment is of some concern. It may indicate that the technicians are simply throwing parts at a problem rather than doing proper testing—and, most importantly, they may not be doing a proper analysis of the test results. Fluid power components are extremely expensive, and replacing a good component can cost hundreds or even thousands of unnecessary dollars.

- If the problem seems to be flow related, does it occur in all or just a few specific directional control valve positions?
- Has there been any overheating of a specific component?
- Does the information indicate a pressure problem?
  - Is the machine able to lift or move a load within its rated capacity?
  - If the problem seems to be pressure related, does it occur in all or just a few specific directional control valve positions?

Review sequence 2, "Problems—What can go wrong in a Fluid Power System" on page 171, explains problems that commonly occur in the various components of fluid power systems.

POSSIBLE CORRECTIONS FOR INCORRECT FLOW

When There Is No Flow

- If the pump is not receiving fluid:
  - Fill the reservoir to the proper level.
  - Inspect, clean, or replace dirty filters.

- Inspect, clean, or replace the reservoir breather vent.
- If there is a lift pump, overhaul or replace it.
- If the pump is powered by an electric motor and is not operating:
  - Overhaul the electric motor.
  - Replace the electric motor.
- If the pump-to-drive coupling is sheared:
  - Check for damage to the pump or pump drive.
  - Replace and align the coupling.
- If the pump motor is turning in the wrong direction:
  - Reverse the direction, which may involve reversing the electrical leads of the motor if the pump is powered by a DC motor.
- If the directional control seems to be set in the wrong position:
  - Check the position of the manually operated controls.
  - Check the electrical circuit on the solenoid-operated controls.
  - Repair or replace the pilot pressure pump if there is one.
- If the entire flow seems to be passing through the relief valve:
  - Adjust the pressure setting on the relief valve.
- If it appears that the pump is damaged:
  - Check for a damaged pump or pump drive.
  - Replace the pump and align the coupling.
- If it appears that the pump is improperly assembled:
  - Overhaul or replace the pump.

When There Is Low Flow

- The flow control may be set too low:
  - Adjust the flow control.
- The pressure relief or unloading valve may be set too low:
  - Adjust or replace the pressure relief valve.
  - Adjust or replace the unloading valve.
- Flow is bypassing through a partially open valve:
  - Check the position of the manually operated controls.
  - Check the electrical circuit on the solenoid-operated controls.
  - Repair or replace the pilot pressure pump if there is one.
  - Overhaul or replace the valve.

- External leaks exist in the system:
  - Tighten leaking connections.
  - Replace leaking hoses.
  - Replace leaking lines.
- The yoke actuating device is inoperative (variable displacement pumps):
  - Overhaul the variable displacement pump.
  - Replace the variable displacement pump.
- The RPM of the pump drive motor is incorrect:
  - Adjust PTO engaged in some speed.
  - Adjust the engine operating parameters.
  - If the drive motor is electric, replace it.
- A worn pump, valve, motor, or cylinder is detected:
  - Overhaul or replace the pump.
  - Overhaul or replace the valve.
  - Overhaul or replace the motor.
  - Overhaul or replace the cylinder.

When There Is Excessive Flow

- The flow control valve is set too high:
  - Adjust the flow control valve.
  - Replace the flow control valve.
- The yoke actuating device is inoperative (variable displacement pumps):
  - Overhaul the variable displacement pump.
  - Replace the variable displacement pump.
- The RPM of the pump drive motor is incorrect:
  - Adjust PTO engaged initial speed.
  - Adjust the engine operating parameters.
  - If the drive motor is electric, replace it.
- An improper pump size has been used as a replacement:
  - Replace it with the correct unit.

POSSIBLE CORRECTIONS FOR INCORRECT PRESSURE

When There Is No Pressure

- Refer to the incorrect flow information earlier.

When There Is Low Pressure

- Refer to the no flow information earlier.
- Refer to the low flow information earlier.
- The pressure reducing valve is set too low:
  - Check the condition of the system fluid.
  - Check the condition of the system filters.
  - Clean the pressure reducing valve.

- Adjust the pressure adjusting valve.
- Replace the pressure adjusting valve.
- Damaged pump, motor, or cylinder:
  - Overhaul, repair, or replace the damaged pump.
  - Overhaul, repair, or replace the damaged motor.
  - Overhaul, repair, or replace the damaged cylinder.

When There Is Erratic Pressure

- Air is in the fluid:
  - Tighten any leaking connections.
  - Fill the reservoir to the proper level.
  - Bleed the air from the system.
- The system has a worn relief valve:
  - Overhaul the relief valve.
  - Replace the relief valve.
- Contamination of the fluid:
  - Replace the dirty filters.
  - Replace the contaminated system fluid.
- The accumulator is defective or has lost its charge:
  - Repair the defective accumulator.
  - Replace the defective accumulator.
- The system has a worn pump, motor, or cylinder:
  - Repair or replace the worn pump.
  - Repair or replace the worn motor.
  - Repair or replace the worn cylinder.

When There Is Excessive Pressure

- A pressure reducing, relief, or unloading valve is misadjusted:
  - Check the condition of the system fluid.
  - Check the condition of the system filters.
  - Clean and adjust the relief or unloading valve.
  - Replace the relief or unloading valve.
- The yoke actuating device is inoperative (variable displacement pumps):
  - Overhaul the variable displacement pump.
  - Replace the variable displacement pump.
- The pressure reducing, relief, or unloading valve is worn or damaged:
  - Repair the worn or damaged valve.
  - Replace the worn or damaged valve.

REVIEW SEQUENCE 2 — Problems—What Can Go Wrong in a Fluid Power System

Although great effort usually goes into designing a system that will operate as problem-free as possible, design errors do occur. More common are the redesign errors that happen when the system is actually installed on a motor vehicle. Still more redesign errors occur during the maintenance and repair of systems in service. Problems are often created by technicians affecting repairs using the wrong hoses, adding couplers at critical flow points, installing lines with angles and bends at critical flow points, and installing elbow fittings. All of these can decrease flow in the system and increase heat. This chapter will look at problems that can occur as a result of system deterioration, poor maintenance practices, and inappropriate repairs.

Engineers often use the term *failure mode* when referring to the things that can go wrong in any engineered system. The typical technician, when they find the source of a problem, such as a hose that is leaking, will make the repair and consider themselves done. The better technician will find the source of the problem, make the repair, and then try to figure out what caused the failure. In fact, the superior technician will try to figure out what caused the failure and make a repair that improves the design and eliminates the root cause of the problem. The superior technician must also keep in mind that any time they make a significant change to the design of the system, the system's designer or a professional engineer should approve the change.

FRICTION OF AIR IN THE HOSE

Most people know that as a spacecraft returns to earth through the atmosphere, the outer skin of the spacecraft becomes very hot. This high temperature is created by the craft falling through the molecules of air in first the upper and the lower atmosphere. At the high speeds of a descending spacecraft, the air generates a great deal of friction along the outer hull of the spacecraft. The same is true of air as it passes through a hydraulic system. It has already been discussed in previous chapters that the hydraulic fluid itself generates friction as it passes through the system. This is what causes the normal heating that occurs during the operation of a fluid power system. Air lacks the lubricating properties of hydraulic fluid, the properties that tend to reduce friction as the fluid passes through the system. Therefore, if air passes through the system, the lack

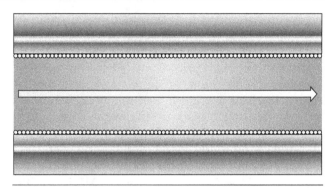

RS 2-1 As the hydraulic fluid moves through the hoses and lines, it is faced with a resistance to flow caused by the walls of the line through that the fluid is moving. This flow generates heat which can cause deterioration in the operation of the system. Hoses that are in good condition and of the proper type and design can reduce the heat generated from the motion of the fluid of the hose. When air is introduced to the hydraulic fluid, the air tends to congregate along the walls of the hose. This adds to the effect of the friction since the air is not as good a lubricant as the oil.

of lubrication coupled with the friction of the air itself moving through the system causes an exceptional amount of heat to build up.

Air can get into a fluid power system through several means. Perhaps the most common ways air gets into the system is as a result of a low level in the reservoir. This is especially true on mobile equipment fluid power applications. As the vehicle starts and stops, turns left and right, and pitches and yaws over terrain, the pickup that supplies hydraulic fluid to the pump will pick up air as the fluid moves away from the pickup. Once the air is in the system, it must pass all the way through the system before eventually returning to the reservoir. Depending on the complexity of the fluid power system, it may take several minutes, hours, or days for the air to be purged thoroughly.

Another more complex and less obvious way for air to enter a hydraulic system is by means of the Bernoulli effect. As the hydraulic fluid flows through the system, the pressure changes. In the areas where the pressure is the lowest and is accompanied by a high velocity, such as in the return line or in a drain line, the pressure in that line can actually drop below atmospheric pressure. At that point, the atmosphere

REVIEW SEQUENCE 2 Problems—What Can Go Wrong in a Fluid Power System (Continued)

can push air into the line. The fluid carrying that air will take it back to the reservoir. If this system is a relatively large system compared to the size of the reservoir, then the fluid will not sit in the reservoir long enough for the air to be released from the hydraulic fluid. As a result, it will simply pass through the system again. This process can take place over and over again until the system shuts down and the fluid is allowed to rest in the reservoir. However, any fluid that is in the hoses, pumps, valves, or actuators of the system will not allow the air to be released and soon the air will become captured or entrained.

Testing

Testing for entrained air in the system is very simple. Simply drain off the sample and take a look. If the fluid seems to have bubbles in it or seems murky and whitish, there is a high probability that the oil is entrained with air. A more sophisticated method for determining this involves a tool that is designed to measure the compressibility of the fluid. Known as an aeration measurement device, it consists of a piston in a cylinder with a hand crank that operates on the piston to compress the fluid. Since hydraulic fluid typically has a compressibility factor of one half of 1 percent by volume, if the fluid being tested has a greater compressibility factor, it is because of air in the fluid. The more compressible the oil sample is, the more air is in the sample.

AIR FLOW AND VACUUM FLOW THROUGH ORIFICES

It may seem obvious, but air and a void (vacuum) each have a different resistance to flow through a hydraulic system than hydraulic oil. This difference becomes most apparent when the air or void passes through an orifice. As the air passes through the orifice, the flow velocity tends to increase. When the oil passes through the orifice, it also increases the flow velocity. This results in a pulsating, or staccato, operation of the hydraulic cylinder or motor. At the same time, the friction of the air passing through the orifice generates heat.

Aerated oil can also cause this effect since the air entrained in the oil causes the density of the oil to be less. The lower-density oil passes more evenly through the orifice. Should the air be evenly disbursed throughout the oil, the pulsation effect will

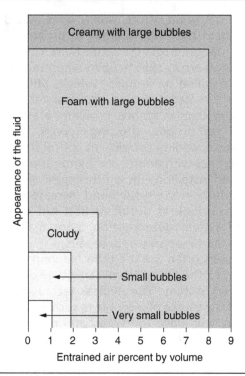

RS 2-2 Any amount of air entrainment can have a negative impact on the operation of a fluid power system. Increased heat, decreased capability, and noise can all result from air in the system. This chart shows the amount of air entrained in a hydraulic fluid based on its appearance. When the oil looks like a light brown cream, then the air content is extremely high. Even the presence of very small bubbles, indicating air content of less than 1 percent, can have an impact. The only true fix for air entrainment is to replace the hydraulic fluid and make sure the system is purged by removing and replacing all the hydraulic fluid in the system.

not be as noticeable, but the heating effect on the oil will increase. Remember that this heating effect is caused by the high friction of the air, the higher velocity, and also by the pressure in the system compressing the air and thereby generating heat.

CAVITATION

Cavitation occurs when there is an interruption in the continuous flow of hydraulic fluid through the system by bubbles of vapor or gas. Vapor occurs

REVIEW SEQUENCE 2
Problems—What Can Go Wrong in a Fluid Power System (Continued)

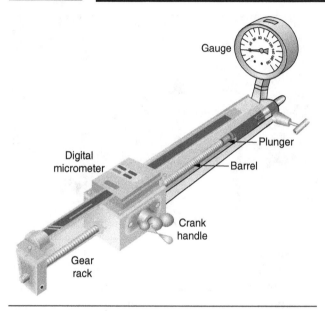

Gauge

Digital micrometer

Plunger

Barrel

Crank handle

Gear rack

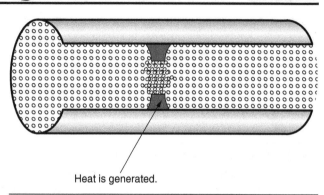

Heat is generated.

RS 2-3 A company called FES has developed a device for taking more accurate readings of air entrainment. Devices such as this can give a very accurate indication of the amount of air in hydraulic fluid. This information can be extremely useful when trying to make a decision concerning whether or not hydraulic oil should be replaced in a large system. Most mobile equipment systems do not fall into the category of large systems. Therefore, except in a large fleet environment with several dozen pieces of mobile hydraulic vehicles, purchasing such a device may not be cost-effective. Observation of the coloration and consistency of the hydraulic oil is usually adequate.

when the pressure of the hydraulic fluid drops below the vaporization point. As the fluid flows through the system and picks up heat, the vaporization pressure increases. This means that the fluid will turn to a vapor at higher and higher pressures. At 20°C (just under 70°F), the pressure of vaporization is very low, about 0.001 atmospheres. Even with very high fluid velocities, this low pressure would be hard to attain. As the temperature rises, the vaporization pressure rises. Exactly how much the vaporization pressure rises will vary depending on the type of hydraulic fluid and the additives being used. As the vaporization pressure increases, it becomes easier for the fluid to vaporize. This increases the opportunity for cavitation to occur.

RS 2-4 There are many places in a fluid power system where the fluid is forced through an orifice. This occurs in some hoses, and many of the valves, and any place where an orifice device has been added to control or limit the flow rate. When the hydraulic fluid gets bubbles in it, these bubbles present an increase in friction as the fluid flows through the various orifices. Two types of bubbles can be present in the hydraulic fluid. The first is air, which seems natural and expected. The second source of bubbles can be the absence of air—meaning the absence of fluid, in fact the absence of anything at all; in other words, bubbles of vacuum passing through the system. Both rob the oil of its lubricating properties and increase the resistance to flow.

The vapor that is formed by cavitation generates a bubble between the fluid and the walls or surface of the piping, pump gears, and other components. Almost as soon as the bubble has formed, its walls begin to weaken. Since this weakening is not even across the surface of the bubble, the surface will break at a single point, allowing fluid to jet through the rupture. This causes a high-pressure stream to "jet" into the component surfaces, etching the surface. The end result is similar to a technology that uses high-pressure water to cut metal. The metal thus becomes damaged.

Damage due to cavitation can be prevented by not altering the operational speed of a system, by changing the hydraulic fluid in accordance with the manufacturer's guidelines, and by using recommended additives. This tendency relates back to Bernoulli's law. Since bends and junctions in a hydraulic system cause a non-laminar (or uneven) flow, the speed of the fluid is also uneven. When the speed of the system is increased, the increased flow

REVIEW SEQUENCE 2 Problems—What Can Go Wrong in a Fluid Power System (Continued)

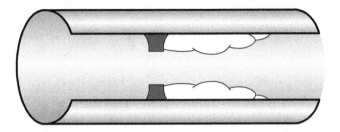

RS 2-5 Cavitation occurs when the flow of fluid through an orifice or other restriction creates a bubble or pocket of low-density fluid. This low-density area is a partial vacuum, which is occupied by no more than a mist or vapor of the fluid. At some point, whether by means of a change in fluid velocity or by a change in pressure, it will collapse. As it collapses, the fluid rushes in to fill the void. This causes the hydraulic fluid to leave behind a mark, or etch the material into which the fluid is impacting. The end result can be very expensive damage. Altering the velocity of the fluid or altering its pressure can reduce the tendency to cavitate.

rate can cause low-pressure areas to be created that would not ordinarily be there. This can promote air bubbles and vapor bubbles, which in turn will lead to cavitation, which in turn can cause damage to components, fittings, and hoses.

Testing

Since cavitation occurs as a result of aeration, it is one piece of evidence that the hydraulic oil may indeed be aerated. There are two primary ways to test for aeration—these are not scientific tests involving highly specialized equipment, but are tests that involve the use of two of the service technician's most effective diagnostic aids. When cavitation occurs, it usually occurs at the point of restriction to flow, such as a fitting, the transition from one host to another, entry into a cylinder or motor, or when passing through an orifice. The movement of air through these point restrictions creates a vibration that may or may not be audible. High-frequency vibration can be felt, while lower frequency can be heard. A heat gun can also be employed at point restrictions through the system to look for a rise in temperature. A moderate rise in temperature is normal—therefore, the technician should be looking for significant increases in temperature.

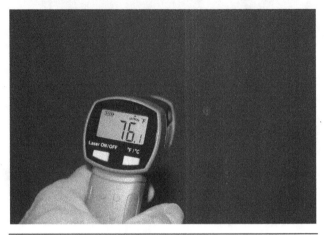

RS 2-6 Heat guns have been around for a couple of decades. They measure infrared in order to remotely detect an object's surface temperature. They are very effective at determining heat being generated within a hydraulic system, assuming the component being tested has metal insulation. These devices once cost several hundred dollars, but today heat guns with a laser sighting device can be purchased for under $100.

LEAKS

The most common causes of leaks are oxidation, heat, overpressurization, and poor maintenance or repairs. Hoses are made of both natural and synthetic compounds that are designed to be flexible and durable. Environmental issues such as exposure to solvents and acids can decrease both their flexibility and durability. Often leaks caused by environmental issues will first be known as the form of an oily area around the fitting or around the dirty area on the hose.

External leaks are very easy to detect; however, the root cause of the leak may require some deduction. Look at the environment, sources of heat, corrosive materials, and general dirt. Once these have been evaluated and determined not to be the root cause, look at the working conditions of the leaking hose. Is the hose being flexed to too tight of an angle? When the hose is extended to its straightest angle, is the hose being overextended and stretched?

Testing

Not surprisingly, testing for a leak in the system usually involves the sense of sight. Often though,

REVIEW SEQUENCE 2 — Problems—What Can Go Wrong in a Fluid Power System (Continued)

RS 2-7 A long-running debate exists about how much seepage can be issued from a hose before it is referred to as a "leak." In this photograph, the hose on the right has an obvious leak as it approaches the upper fitting. The question for the technician is: Does the dampness on the hose on the left originate with the leak or is it from the hose on the right? It would appear that the leak originates where the metal joins the rubber part of the hose. Although it might be tempting to simply trim the hose crimp on a new fitting, to do so might create more problems. First, the hose is partially deteriorated and therefore the likelihood of a good seal between the new fitting and the hose is unlikely, no matter how small the current is. The other is that shortening the hose would decrease the length of the hose and therefore decrease the size of the articulation loop.

leaks are small and will not be witnessed as a steady stream of fluid exiting the system. Also, leaks can occur at only specific times during the operation of the system. In most systems, not all hoses are pressurized at all times. Therefore, it is important to observe the system for leaks during the entire operational cycle of the system. In stationary hydraulic systems, a puddle of oil, or an oil stain, is a good indication that the leak is in the immediate proximity. In mobile equipment, the truck or machine can at times be moving down the road at nearly 70 miles an hour. This can create wind currents and eddies well over 100 miles an hour. This means that the oil puddle or stain can

present itself many feet from the actual location of the leak. The best procedure is to ask a co-worker to operate the equipment in all of its modes while observing all the hoses, fittings, connections, and components.

For smaller leaks, it is unlikely that a puddle or stain will be found all. Here the best evidence would be the presence of an oily looking film or dirt near the point of the leak.

HOSE RESTRICTIONS

Kinks

Three types of problems can restrict the flow through a hose. Kinks relate to hose routing. Care needs to be taken when installing a hose on an actuator to make sure the hose does not form a kink throughout the range of motion of that actuator. If the hose carrying the hydraulic fluid to the actuator doubles back on itself over a tight radius, a kink can occur. This can create a restriction to flow. When new hoses are installed and old hoses are reinstalled, they should have enough slack that the hose is able to form a large loop rather than kinking.

RS 2-8 There is a very real risk, especially with short hoses, that improper installation can cause a case where it will restrict fluid flow. This cannot only reduce the speed at which equipment operates, it can also generate excessive heat. When replacing a hose, it is important to have the proper material, the proper fitting, and the proper diameter.

REVIEW SEQUENCE 2

Problems—What Can Go Wrong in a Fluid Power System (Continued)

Testing

As with leaks, the best way to test for kinking hoses, which can intermittently affect the operation of a hydraulic system, is to have a co-worker operate the equipment throughout its complete range, while observing the flexible lines. It is important to keep in mind that even the best-explained symptoms can leave out key factors about when the problem occurs. For instance, a system operator might inform the technician that an actuator operate slowly. The observant technician might find that the actuator, such as a cylinder, operates at normal speed and without vibration throughout the first half of the cylinder stroke. The second half of the stroke may cause a hose to kink, thereby slowing the operation, or cause a vibration, but only in the second half of the travel for the piston. Although gathering information about symptoms from the operator is critical to making a proper diagnosis, a good technician will always keep in mind that operator explanations are not always thorough.

CONTAMINATION

Contamination can cause a restriction in a hose or line in two ways. The first is for a contaminating particle to wedge itself in the channel of flow. How this impedes flow is obvious. Other, more subtle, forms of contamination can affect the flow of hydraulic fluid through a hose. These come in the form of substances that are either held in suspension in the hydraulic fluid or are dissolved in the fluid. These substances can be either particulate or liquid. Carried by the hydraulic fluid, they abrade the interior of the hose and cause it to collapse internally, or they break down the interior chemically, thereby weakening the internal structure of the hose. The best line of defense against restrictions due to contamination is to change the fluid and filters according to the system manufacturer's recommendations.

Testing

Testing for some contamination is quite easy. The technician will find shiny fragments in the oil, or when a sample of the oil is rubbed between the thumb and forefinger it will have a gritty feel. Hard-particle contamination, therefore, is often self-evident. Contaminating liquids and gases, however,

RS 2-9 Contamination can be extremely harmful to any fluid power system. Fluid power systems operate as efficiently as they do because of very closely tolerant components. Small pieces of grit, dirt, and other foreign material can severely damage or destroy these closely tolerant components. This photo shows an example of how contaminants building up behind the piston inside of the cylinder have slowly eroded and destroyed the cylinder at each end.

are far less obvious. Many organizations will regularly send samples of hydraulic oil for their equipment to an oil analysis laboratory. This can be expensive.

Today, many organizations will only send oil out for analysis when there have been multiple failures either in the hard parts of the system, such as pumps and actuators, or in the soft components, like seals and hoses. In a large system, where many gallons of fluid would need to be replaced, regular oil analysis makes sense. In smaller systems where contamination is suspected, the oil should simply be purged from the system, following the manufacturer's guidelines, and then fresh oil should be installed.

INTERNAL COLLAPSE

Internal collapse is caused by deterioration in the hose. This might be because of hard-particle abrasive damage, because of degrading of the material the hose is made from, or because of a chemical reaction between the inner surface of the hose and

REVIEW SEQUENCE 2 — Problems—What Can Go Wrong in a Fluid Power System (Continued)

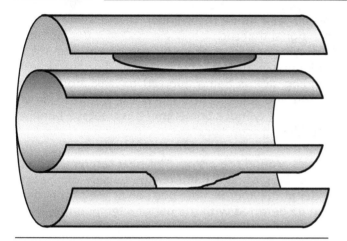

RS 2-10 Two more difficult problems to diagnose concern a hydraulic hose that has an internal embolism or a tear causing a restriction. With the embolism, fluid leaks into the layers of the hose through a perforation or tear in the tube of the hose. The fluid trapped between the two reinforcing layers causes the tube to bulge inward on the passageway through the hose. This restricts the flow of fluid through the hose and eventually can tear it. Once the tear takes place, the inner tube can be forced downward, almost completely blocking the passageway through the hose. What makes each of these difficult to troubleshoot is that the restriction is not always there. Sometimes when the machinery is operating the embolism or tear adheres to the reinforcing layers and there is no restriction to flow.

any contaminants being carried by the hydraulic fluid. While these are things that can accelerate deterioration of the hoses in a hydraulic system, oxidation and deterioration of the hoses can simply be the result of aging.

Testing

The internal collapse of hydraulic hoses is rarely evident visually. A symptom of this will almost always be the slow or erratic operation of the actuator. The operator may complain that the actuator seems to operate at normal speed sometimes or even most of the time. In many cases, this complaint alone may be enough to make a diagnosis. If several hoses are involved, then it may be necessary to use a heat gun to look for hotspots along the hoses to or

from the actuator. One could also use a flow meter to bypass the suspected segment of hose. An astute technician will remember that those restrictions can occur on both the pressure side and the return side of the actuator.

WORN PUMPS

Pumps are under continuous assault when the system is operating. This includes assault from pressure, friction, heat, and contaminants. Regardless of the type of pump, all are susceptible to damage and wear from these sources.

A worn pump can be identified in two primary ways. Worn pumps tend to have increased mechanical friction as they operate, the result of which is higher operational temperatures. Worn pumps also tend to bypass hydraulic fluid internally. This bypassing creates friction from the fluid passing across the wrong side of the pumping mechanism and this also increases temperature. These temperatures can be detected with a contact temperature probe or

RS 2-11 Even with the best lubrication pumps in place, hydraulic motors are subject to wear. There are additives in the hydraulic fluid that enhance lubrication to protect bearings surfaces and mating surfaces in both motors and pumps. Worn pumps tend to have both a reduced maximum pressure and maximum flow. This is a result of internal bypassing within the pump. A symptom that might result from a worn pump is a hydraulic system's inability to lift or move a mass that is within the limits of the design of the fluid power–based machine.

Problems—What Can Go Wrong in a Fluid Power System (Continued)

with an infrared "heat gun." The other factor that affects the operation and efficiency of a pump, and therefore the rest of the hydraulic system, is that the worn hydraulic pump can bypass fluid internally and therefore have a decreased flow rate and a decreased maximum pressure.

Testing

The first symptom of a worn pump is always the inability to flow the proper amount of hydraulic fluid. Therefore, when an operator complains of the larger actuators of the equipment operating slowly, the technician should test the maximum flow rate of the pump. This is done by connecting a flow meter to the outbound side of the pump and routing the flow meter directly back to the reservoir. At this point, the maximum flow through the pump will occur. The rate of flow should be checked against manufacturer specifications for the pump. If these specifications are not available from the equipment manufacturer, they should be available from the pump manufacturer.

Should the equivalent operator not notice that the larger actuators have begun to operate slowly, eventually the pump will become warm enough that he cannot push against resistance to flow to create the proper pressure. The operator would indicate that the equipment appeared to be "weak." The technician can troubleshoot this by loading the equipment to about 90 percent of its rated capacity, have someone operate the equipment, and look for high temperatures. Under these conditions, temperatures will climb throughout the entire system as a result of pressures being at near maximum. If the system is able to manipulate the load up to 90 percent of the rated capacity, then by definition the system was not weak. However, if the operator is correct, the system will begin to bypass internally. The point of bypassing, such as a pressure relief valve, or in this case the pump, will climb in temperature.

LEAKING CYLINDER SEALS

Leaking cylinder seals come in two forms. The first is the obvious external leak. Generally speaking, this type of leaking cylinder seal manifests itself as a loss of hydraulic fluid around the shaft of the rod attached to the piston. A much more difficult leaking seal to spot is one that occurs around one of the

piston seals inside the cylinder. Several telltale signs can signal that such a leak exists. One is the fact that the cylinder will appear to "leak down" when it is holding a load. This is because the oil on the closed, holding side is being allowed to leak around the piston and seals, which allows the piston to move within the cylinder. Sometimes this creates a familiar shuddering sound and motion. Another symptom of an internal leak is a buildup of heat as the fluid moves around the piston. This heat can be detected either by placing a hand on the outside of the cylinder at the approximate position of the piston, or it can be detected with the use of a temperature gun.

Testing

Leaking cylinder seals can often be difficult to detect. In many cases, the cylinder walls are very thick and slow to transfer heat. The technician can verify leaking cylinder seals by connecting a flow meter to the return line and parking the cylinder in the mid

RS 2-12 Over an extended period of time, the seals and most of the cylinder will deteriorate. The life expectancy of the seals can be maximized through good filtration and routine replacement of the hydraulic fluid. Like with most components in a fluid power system, the most significant cause of failure for piston seals in a hydraulic cylinder is contamination. Unlike most other components, rapid deterioration of the piston seals can result from exposure to incorrect fluids. So in addition to hard-particle contamination, the hydraulic fluid should be inspected for water contamination and accidental contamination by other fluids such as diesel fuel or gasoline.

REVIEW SEQUENCE 2

Problems—What Can Go Wrong in a Fluid Power System (Continued)

range with approximately a 90 percent load. If the cylinder seals are leaking, oil will be transferring from one side of the piston to the other side. The piston will then slide back into the cylinder. This will generate heat and often vibration. When a technician notices that the piston is beginning to retract under the 90 percent load, he or she should then observe the flow meter connected in the return line. If it indicates flow, then chances are the cylinder seals are good and the directional control valve or other component is causing the leak down. Technicians should keep in mind that any part of a cylinder that is slow in retracting as a result of a leak outside of the cylinder will have a very small flow rate back to the reservoir. Many flow meters are not capable of reading very small flow rates. It may be necessary to open the return line and allow the fluid escaping past the directional control valve or other leaking component to flow into a safety-authorized container.

Always remember when load testing equipment to keep personnel and other equipment away from the equipment being tested since a retracting piston cannot allow booms and other devices to cause damage if they are in the way.

WORN MOTORS

A worn hydraulic motor is much like a worn pump. As the mechanical internal parts of the pump wear, the clearances between the components increase, and this allows fluid to bypass the pumping mechanism. The result is a poorly performing motor. The motor may fail to hold in a fixed position; it may be weak and unable to lift or move heavy loads; or it may be operating at a reduced efficiency because of the extra heat it is generating. Like worn and internally bypassing pumps, the movement of the motor can exhibit vibration, excessive heat, and an inability to do its job effectively. The cause of worn motors is usually either old age or contamination in the system. The very contaminants that deteriorate the pumps in the system are the contaminants that can and will damage the inner workings of hydraulic motors.

Most problems in fluid power systems result from poor maintenance. In most cases, especially in mobile fluid power systems, the system is built to tolerances that far exceed what the work capacity will be during normal operation. When boiled down to the basics, only a few symptoms are related to problems found in mobile equipment hydraulics.

RS 2-13 Hydraulic motors are subject to the same types of damage and deterioration as hydraulic pumps. What can make this difficult to diagnose is that in a hydraulic motor–based system it is sometimes difficult to determine whether the problem is a worn motor or a worn pump. The symptoms for each are similar. Proper troubleshooting techniques using a flow meter pressure tester can help pinpoint the source of the problem. Isolating and testing first the pump and then the motor can isolate which component is defective.

Overheating throughout the system is usually a result of low oil capacity or a problem with the hydraulic system cooling heat exchanger, if there is one. Overheating within specific components is usually a result of the friction from fluid passing through a small passageway. In the case of an orifice, heat at the orifice is normal. Excessive heat at the orifice could be the result of air in the system, or bubbles passing through the orifice. If the component is not designed to have an orifice, yet there is heating of the component, it is an indication that fluid may be bypassing that component internally. For instance, if the symptom was a cylinder that was unable to be part of the mid range of travel for the piston, there could be many possible causes. One cause might be that the braking valves are leaking, in which case they would be hot since the operator attempted to park the cylinder in the midrange position. It is also possible that fluid is bypassing the seals that not only contribute to developing force on the piston but also separate the extended side of the cylinder from the retract side of the cylinder. This can be detected by using an infrared heat gun

Problems—What Can Go Wrong in a Fluid Power System (Continued)

REVIEW SEQUENCE 2

and measuring the temperature at the piston position while it is holding at mid range, and then comparing that to the temperature of the cylinder ends. A significant difference in temperature points to hydraulic oil bypassing the seals.

Shuddering or vibration in the system is an indication of inconsistency in the character of the fluid. That is to say that the density of the fluid is changing while the system is in operation. This can be caused by the hydraulic fluid being contaminated with an inappropriate fluid, such as water, or it could be an indication that air has been entrained into the hydraulic oil. In either case, an expensive and effective way to deal with the problem is simply to service the equipment by changing the hydraulic oil. If it turns out that air is entrained in the oil, the well will be discolored, looking almost milky, an effort should be made to determine the source of the air.

A slow operating system, one where the hydraulic motor or cylinder moves slower than the rated speed, can be caused by insufficient fluid flow to the cylinder. One of the first things to look at is to determine if the fluid is being routed to the wrong place. Look for places where oil could be bypassing the main path to the actuator and returning to the reservoir prematurely, or bypassing the reservoir and simply leaking on the ground. If there are no internal or external leaks, then there must be something restricting the flow inside the system. This can be caused by a worn pump or a worn actuator, such as a motor or cylinder. Hoses that are kinked or have collapsed internally can cause a reduction in the flow rate of the fluid. Components, such as valves and other controls, which were in service when the system was contaminated, can be restricted as a result of a buildup due to hard-particle material.

Thorough routine maintenance inspections can decrease the likelihood of the system becoming internally damaged or contaminated. It is always easier to follow the manufacturer's recommended service procedures and intervals than it is to diagnose and make repairs.

Testing

Testing for a worn motor is a lot like testing a cylinder. The first thing the operator will complain about is the inability to lift the proper load. Essentially this is due to oil bypassing the pistons or gears of the motor and returning directly to the reservoir. The motor should be loaded to approximately 90 percent of its rated capacity. The technical specifications from the manufacturer should include speed and flow rate information. Connect a flow meter on the inbound line of the motor or on the outbound line if the inbound line is inaccessible. Operate the equipment with a 90-percent load and observe the flow meter. If the flow rate is higher than is indicated by the specifications for the speed at which the motor is being operated, bypassing inside the motor is occurring. The technician may be able to confirm this by the presence of excessive heat on the case of the motor.

If the manufacturer's specifications are incomplete, then the technician can install a flow meter where the hose enters the motor and attempt to park the motor with a 90-percent load. If the motor does not park because of internal leaks outside of the motor, then the flow meter will indicate zero flow. If there is an indication of flow on the meter, the motor is bypassing internally and must be replaced or repaired.

Review Questions

1. The interview process is an important part of the troubleshooting procedure because:

   A. It gets the system operator involved in the troubleshooting process.

   B. It gives the technician specific details about particular concerns the system operator has.

   C. It allows the technician to get a history related to how the system is not functioning.

   D. All of the above

2. In the counterbalance circuit shown in the accompanying figure, the directional control valve has been activated to the straight through flow position, but the load is not lowering. The pressure gauge shows high pressure (above normal) and a flow meter is indicating zero flow. When a technician activates the directional control valve to the straight through flow position, the valve can be heard to change position. There are many possible causes. Of the following possible causes, which is the most likely?

   A. The check valve is leaking.

   B. The counterbalance valve is not opening.

   C. The piston seals are bypassing inside the cylinder.

   D. The system filter is restricted.

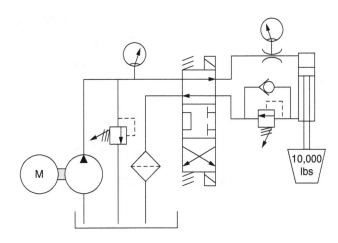

3. On another vehicle, the same counterbalance circuit is allowing the 10,000-pound weight to slowly lower when the directional control valve is in the center position. There is a minimal amount of pressure, only a few PSI, and the line between the pump and the directional control valve. The flow meter is indicating a small amount of flow as the weight slowly descends. Which of the following could not be the cause?

   A. A directional control valve with an internally in the center closed position

   B. A leaking counterbalance valve

   C. A leaking check valve

   D. A leaking filter

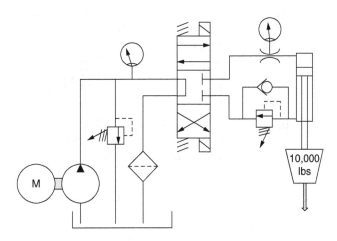

4. In the unloading circuit shown in the accompanying illustration, when the directional control valve is in the straight through flow position, the system is unable to move the weight. Using an infrared heat gun, the technician notes that the temperature of the small pump is over 200°F. The temperature of the large pump, however, is barely above room temperature. There is maximum flow in the return line of the large pump. Of the following items, which component is the most likely cause of the symptoms?

    A. The unloader valve   C. The check valve

    B. The pressure relief valve   D. A defective large pump

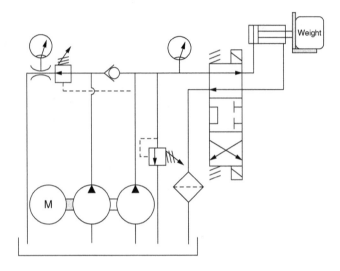

5. On a different vehicle, the same unloading circuit moves the weight with maximum speed and maximum force every time the directional control valve is moved to the straight through flow position. Of the following components, which is the most likely source of the problem?

    A. A defective electric motor   C. An unloading valve stuck in the closed position

    B. A defective directional   D. A defective check valve
       control valve

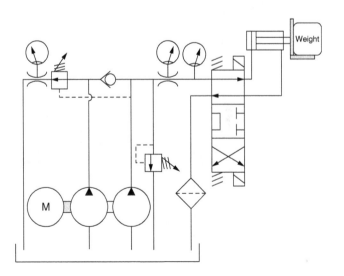

6. In the hydraulic motor circuit shown in the preceding illustration, the directional control valve is moved to the closed center position. The weight suspended from the pulley attached to the motor slowly causes the motor to rotate and the way to spool downward. Also note that there is a small flow through the supply line to the motor when the directional control valve is in the closed center position. This could be caused by any of the following except:

A. The pump is bypassing internally.

B. The motor is bypassing internally.

C. The brake of the valve is defective.

D. There is an external leak between the motor and the brake valve.

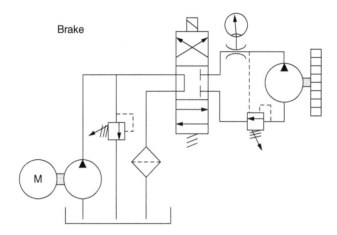

7. The preceding photograph shows a combination flow control and check valve. Assuming the check valve portion is operating normally, which of the following symptoms could be caused by this component?

A. A motor that turns too fast

B. A cylinder piston that slowly descends when parked

C. A cylinder piston that does not move

D. A piston that moves too slow

8. In the fluid power circuit shown here, the directional control valve does not change position. Which of the following tests should be done first?

   A. All fuses associated with the fluid power system should be checked.

   B. The filter should be replaced.

   C. Flow into the retract side of the cylinder should be tested with a flow meter.

   D. The directional control valve should be replaced.

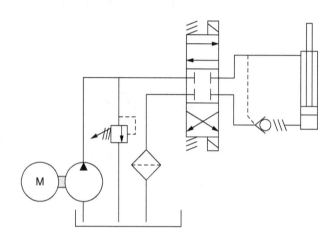

9. Technician A says that the vehicle battery is an important part of the fluid power system in mobile equipment. Technician B says that one of the first checks that should be done for any symptoms is to inspect the fluid level in the reservoir. Who is correct?

   A. Technician A only

   B. Technician B only

   C. Both Technician A and Technician B

   D. Neither Technician A nor Technician B

10. A good fluid power technician will:

    A. Always wear proper personal protection equipment.

    B. Develop a diagnostic plan based on the symptoms before beginning to use tools.

    C. Be able to demand a higher salary than a technician who does not understand and is not able to troubleshoot fluid power systems.

    D. All of the above

11. The root cause of a slow operating cylinder or motor is always related to:

    A. Heat

    B. Air in the system

    C. High flow rate

    D. Low flow rate

12. High flow rates in a system can contribute to any of the following except:

    A. Slow system operation

    B. Cavitation

    C. Vibration

    D. Shuddering

13. Technician A says that when a cylinder fails to hold its position in the mid range, it is always because of bad piston seals. Technician B says that the piston of the cylinder may fail to hold its position in the mid range because of bad piston seals or a defect in the braking mechanism for the cylinder. Who is correct?

    A. Technician A

    B. Technician B

    C. Both Technician A and Technician B

    D. Neither Technician A nor Technician B

14. An embolism on the interior of the hose is:

    A. A restriction caused by hard-particle contamination.

    B. A restriction formed by a bubble between the tube layer and the reinforcing layer of the hose.

    C. A perforation in the hose that causes an excessive flow rate.

    D. A sign that the hose is badly kinked.

15. Technician A says that it is a mistake to install a hose on an articulated component of a hydraulic system that is too short, but it is okay to install one that is too long. Technician B says that it is okay to install a hose that is too short, but it is a mistake to install one that is too long. Who is correct?

    A. Technician A

    B. Technician B

    C. Neither Technician A nor Technician B

    D. Both Technician A and Technician B

16. When a cylinder is disassembled, pitting, scoring, and gouging are found. Which of the following could cause this?

    A. Liquid contamination of the fluid

    B. Hard-particle contamination of the fluid

    C. Additive breakdown of the fluid

    D. All of the above

17. High heat at one particular point in the system while the rest of the system remains at normal temperature can indicate:

    A. The wrong fluid has been used.

    B. The system is operating at excessively high pressure.

    C. There is a restriction at that point in the system.

    D. Nothing. This should be considered normal.

18. A general temperature increase in the system is most likely caused by:

    A. Air entrainment in the system oil.

    B. A low fluid level.

    C. An ineffective cooling heat exchanger.

    D. Any of the above

19. Cavitation is:

    A. When the pump is sucking air from the reservoir.

    B. Air or vacuum bubbles forming in the system and then being collapsed in a pressurized fluid.

    C. A leak in the reservoir.

    D. A condition that causes the slow operation of pumps and cylinders but otherwise has no impact on the condition or life expectancy of the system.

20. When troubleshooting a fluid power system, the condition of the oil should never be taken into account since it is not a variable in the operation of the system.

    A. True

    B. False

APPENDIX

Conversions

CONVERSION TABLE FOR UNITS OF MEASURE

Atmospheres	Feet of Water	33.9
Atmospheres	Inches of Mercury (Hg)	29.92
Atmospheres	PSI (Pounds per Sq. Inch)	14.7
BTU	Foot Pounds	778.3
BTU per Hour	Watts	0.2931
BTU per Minute	Horsepower	0.02356
Celsius (Centigrade)	Fahrenheit	°C × 1.8 + 32
Centimeters	Inches	0.3937
Cubic Centimeters	Gallons (U.S. Liquid)	0.0002642
Cubic Centimeters	Liters	0.001
Cubic Feet	Cubic Inches	1,728
Cubic Feet	Gallons (U.S. Liquid)	7.48052
Cubic Inches	Cubic Feet	0.0005787
Cubic Inches	Gallons (U.S. Liquid)	0.004329
Days	Seconds	86,400
Degrees (Angle)	Radians	0.01745
Feet	Meters	0.3048
Feet	Miles	0.0001894
Feet of Water	Atmospheres	0.0295
Feet of Water	Inches of Mercury (Hg)	0.8826
Feet of Water	PSI (Pounds per Sq. Inch)	0.4335
Feet per Minute	Miles per Hour	0.01136
Feet per Second	Miles per Hour	0.6818
Foot-Pounds	BTU	0.001286
Foot-Pounds per Minute	Horsepower	0.0000303
Foot-Pounds per Second	Horsepower	0.001818
Gallons U.S. Liquid	Cubic Feet	0.1337
Gallons U.S. Liquid	Cubic Inches	231
Gallons of Water	Pounds of Water	8.3453
Horsepower	BTU per Minute	42.44
Horsepower	Foot-Pounds per Minute	33,000
Horsepower	Foot-Pounds per Second	550
Horsepower	Watts	745.7
Hours	Days	0.04167
Hours	Weeks	0.005952

(continued)

CONVERSION TABLE FOR UNITS OF MEASURE (*continued*)

Inches	Centimeters	2.54
Inches of Mercury (Hg)	Atmospheres	0.03342
Inches of Mercury (Hg)	Feet of Water	1.133
Inches of Mercury (Hg)	PSI pounds per Sq. Inch	0.4912
Inches of Water	PSI Pounds per Sq. Inch	0.03613
Liters	Cubic Centimeters	1,000
Liters	Gallons (U.S. Liquid)	0.2642
Micron	Inches	0.00004
Miles (Statute)	Feet	5,280
Miles per Hour (M.P.H.)	Feet per Minute	88
Miles per Hour	Feet per Second	1.467
Ounces (Weight)	Pounds	0.0625
Ounces (Liquid)	Cubic Inches	1.805
Pints (Liquid)	Quarts (Liquid)	0.5
Pounds	Grains	7,000
Pounds	Grams	453.59
Pounds	Ounces	16
PSI (Pounds per Sq. Inch)	Atmospheres	0.06804
PSI (Pounds per Sq. Inch)	Feet of Water	2.307
PSI (Pounds per Sq. Inch)	Inches of Mercury (Hg)	2.036
Quarts	Gallons	0.25
Square Feet	Square Inches	144
Temperature $(°F) = (°C) \times 9/5 + 32$	Temperature $(°C) = ([°F] - 32) \times 5/9$	
Tons (U.S.)	Pounds	2,000
Watts	Horsepower	0.001341

FRACTIONAL EQUIVALENTS IN DECIMAL AND METRIC

64ths	32nds	16ths	8ths	Decimal	mm
1/64				0.01562	0.397
	1/32			0.03125	0.794
3/64				0.04688	1.191
		1/16		0.06250	1.588
5/64				0.07812	1.984
	3/32			0.09375	2.381
7/64				0.10938	2.778
			1/8	0.12500	3.175
9/64				0.14062	3.572
	5/32			0.15625	3.969
11/64				0.17188	4.366
		3/16		0.18750	4.763
13/64				0.20312	5.159
	7/32			0.21875	5.556
15/64				0.23438	5.953
			1/4	0.25000	6.350

FRACTIONAL EQUIVALENTS IN DECIMAL AND METRIC (*continued*)

64ths	32nds	16ths	8ths	Decimal	mm
17/64				0.26562	6.747
	9/32			0.28125	7.144
19/64				0.29688	7.541
		5/16		0.31250	7.938
21/64				0.32812	8.334
	11/32			0.34375	8.731
23/64				0.35938	9.128
			3/8	0.37500	9.525
25/64				0.39062	9.922
	13/32			0.40625	10.319
27/64				0.42188	10.716
		7/16		0.43750	11.113
29/64				0.45312	11.509
	15/32			0.46875	11.906
31/64				0.48438	12.303
			1/2	0.50000	12.700
33/64				0.51562	13.097
	17/32			0.53125	13.494
35/64				0.54688	13.891
		9/16		0.56250	14.288
37/64				0.57812	14.684
	19/32			0.59375	15.081
39/64				0.60938	15.478
			5/8	0.62500	15.875
41/64				0.64062	16.272
	21/32			0.65625	16.669
43/64				0.67188	17.066
		11/16		0.68750	17.463
45/64				0.70312	17.859
	23/32			0.71875	18.256
47/64				0.73438	18.653
			3/4	0.75000	19.050
49/64				0.76562	19.447
	25/32			0.78125	19.844
51/64				0.79688	20.241
		13/16		0.81250	20.638
53/64				0.82812	21.034
	27/32			0.84375	21.431
55/64				0.85938	21.828
			7/8	0.87500	22.225
57/64				0.89062	22.622
	29/32			0.90625	23.019
59/64				0.92188	23.416
		15/16		0.93750	23.813

(*continued*)

FRACTIONAL EQUIVALENTS IN DECIMAL AND METRIC (*continued*)

64ths	32nds	16ths	8ths	Decimal	mm
61/64				0.95312	24.209
	31/32			0.96875	24.606
63/64				0.98438	25.003
1	1	1	1	1.00000	25.400

TEMPERATURE CONVERSION

°C	°F	°C	°F	°C	°F	°C	°F	°C	°F
−40	−40	−4	24.8	32	89.6	68	154.4	104	219.2
−39	−38.2	−3	26.6	33	91.4	69	156.2	105	221
−38	−36.4	−2	28.4	34	93.2	70	158	106	222.8
−37	−34.6	−1	30.2	35	95	71	159.8	107	224.6
−36	−32.8	0	32	36	96.8	72	161.6	108	226.4
−35	−31	1	33.8	37	98.6	73	163.4	109	228.2
−34	−29.2	2	35.6	38	100.4	74	165.2	110	230
−33	−27.4	3	37.4	39	102.2	75	167	111	231.8
−32	−25.6	4	39.2	40	104	76	168.8	112	233.6
−31	−23.8	5	41	41	105.8	77	170.6	113	235.4
−30	−22	6	42.8	42	107.6	78	172.4	114	237.2
−29	−20.2	7	44.6	43	109.4	79	174.2	115	239
−28	−18.4	8	46.4	44	111.2	80	176	116	240.8
−27	−16.6	9	48.2	45	113	81	177.8	117	242.6
−26	−14.8	10	50	46	114.8	82	179.6	118	244.4
−25	−13	11	51.8	47	116.6	83	181.4	119	246.2
−24	−11.2	12	53.6	48	118.4	84	183.2	120	248
−23	−9.4	13	55.4	49	120.2	85	185	121	249.8
−22	−7.6	14	57.2	50	122	86	186.8	122	251.6
−21	−5.8	15	59	51	123.8	87	188.6	123	253.4
−20	−4	16	60.8	52	125.6	88	190.4	124	255.2
−19	−2.2	17	62.6	53	127.4	89	192.2	125	257
−18	−0.4	18	64.4	54	129.2	90	194	126	258.8
−17	1.4	19	66.2	55	131	91	195.8	127	260.6
−16	3.2	20	68	56	132.8	92	197.6	128	262.4
−15	5	21	69.8	57	134.6	93	199.4	129	264.2
−14	6.8	22	71.6	58	136.4	94	201.2	130	266
−13	8.6	23	73.4	59	138.2	95	203	131	267.8
−12	10.4	24	75.2	60	140	96	204.8	132	269.6
−11	12.2	25	77	61	141.8	97	206.6	133	271.4
−10	14	26	78.8	62	143.6	98	208.4	134	273.2
−9	15.8	27	80.6	63	145.4	99	210.2	135	275
−8	17.6	28	82.4	64	147.2	100	212	136	276.8
−7	19.4	29	84.2	65	149	101	213.8	137	278.6
−6	21.2	30	86	66	150.8	102	215.6	138	280.4
−5	23	31	87.8	67	152.6	103	217.4	139	282.2

WIRE GAUGE INFORMATION AND EQUIVALENTS (ONLY COMMON WIRE SIZES ARE LISTED)

American Wire Gauge (AWG)	Diameter (mm)	Diameter (in)	Cross-sectional Area (mm²)	Resistance (ohm/1,000 m)
40	0.08	0.0032	0.0050	3,420
39	0.09	0.0035	0.0064	2,700
38	0.10	0.0040	0.0080	2,190
37	0.11	0.0045	0.0100	1,810
36	0.13	0.005	0.013	1,300
35	0.14	0.0056	0.016	1,120
34	0.16	0.0063	0.020	844
33	0.18	0.0071	0.025	676
32	0.20	0.008	0.032	547
30	0.25	0.01	0.050	351
28	0.33	0.013	0.08	232.0
27	0.36	0.018	0.10	178
26	0.41	0.016	0.13	137
25	0.45	0.018	0.16	108
24	0.51	0.02	0.20	87.5
22	0.64	0.025	0.33	51.7
20	0.81	0.032	0.50	34.1
18	1.02	0.04	0.82	21.9
16	1.29	0.051	1.3	13.0
14	1.63	0.064	2.0	8.54
13	1.80	0.072	2.6	6.76
12	2.05	0.081	3.3	5.4
10	2.59	0.10	5.26	3.4
8	3.25	0.13	8.30	2.2
6	4.115	0.17	13.30	1.5
4	5.189	0.20	21.2	0.8
2	6.543	0.26	33.6	0.5
1	7.348	0.29	42.4	0.4
0	8.252	0.33	53.5	0.31
00 (2/0)	9.266	0.37	67.4	0.25
000 (3/0)	10.40	0.41	85.0	0.2
0000 (4/0)	11.684	0.46	107.0	0.16

WIRE RECOMMENDATIONS

Circuit Amperes		Circuit Watts		Wire Gauge for Length in Feet						
6 V	12 V	6 V	12 V	3'	5'	7'	10'	15'	20'	25'
0 to 2.5	0 to 5	15	30	18	18	18	18	18	18	18
3.0	6	18	36	18	18	18	18	18	18	16
3.5	7	21	42	18	18	18	18	18	18	16
4.0	8	24	48	18	18	18	18	18	16	16
5.0	10	30	60	18	18	18	18	16	16	16

(continued)

WIRE RECOMMENDATIONS (*continued*)

Circuit Amperes		Circuit Watts		Wire Gauge for Length in Feet						
6 V	12 V	6 V	12 V	3′	5′	7′	10′	15′	20′	25′
5.5	11	33	66	18	18	18	18	16	16	14
6.0	12	36	72	18	18	18	18	16	16	14
7.5	15	45	90	18	18	18	18	14	14	12
9.0	18	54	108	18	18	16	16	14	14	12
10	20	60	120	18	18	16	16	14	12	10
11	22	66	132	18	18	16	16	12	12	10
12	24	72	144	18	18	16	16	12	12	10
15	30	90	180	18	16	16	14	10	10	10
20	40	120	240	18	16	14	12	10	10	8
25	50	150	300	16	14	12	12	10	10	8
50	100	300	600	12	12	10	10	6	6	4
75	150	450	900	10	10	8	8	4	4	2
100	200	600	1,200	10	8	8	6	4	4	2

FORMULAS

Hydraulic Pump Calculations

Horsepower Required to Drive Pump = GPM × PSI × 0.0007 (this is a "rule-of-thumb" calculation)

How many horsepower are needed to drive a 10-gpm pump at 1,750 psi?

GPM = 10

PSI = 1,750

GPM × PSI × 0.0007 = 10 × 1,750 × 0.0007 = 12.25 horsepower

Pump Output Flow (in Gallons Per Minute) = RPM × Pump Displacement/231

How much oil will be produced by a 2.21-cubic-inch pump operating at 1,120 rpm?

RPM = 1,120

Pump Displacement = 2.21 cubic inches

RPM × Pump Displacement/231 = 1,120 × 2.21/231 = 10.72 gpm

Pump Displacement Needed for GPM of Output Flow = 231 × GPM/RPM

What displacement is needed to produce 7 gpm at 1,740 rpm?

GPM = 7

RPM = 1,740

231 × GPM/RPM = 231 × 7/1,740 = 0.93 cubic inches per revolution

Hydraulic Cylinder Calculations

Cylinder Blind End Area (in Square Inches) = Pi × (Cylinder Radius)²

What is the area of a 6-inch-diameter cylinder?

Diameter = 6"
Radius is 1/2 of the diameter = 3"
Radius² = 3" × 3" = 9"
Pi × (Cylinder Radius)² = 3.14 × (3)² = 3.14 × 9 = 28.26 square inches

Cylinder Rod End Area (in Square Inches) = Blind End Area − Rod Area

What is the rod end area of a 6-inch-diameter cylinder that has a 3-inch-diameter rod?

Cylinder Blind End Area = 28.26 square inches
Rod Diameter = 3"
Radius is 1/2 of the rod diameter = 1.5"
Radius² = 1.5" × 1.5" = 2.25"
Pi × Radius² = 3.14 × 2.25 = 7.07 square inches
Blind End Area − Rod Area = 28.26 − 7.07 = 21.19 square inches

Cylinder Output Force (in Pounds) = Pressure (in PSI) × Cylinder Area

What is the push force of a 6-inch-diameter cylinder operating at 2,500 PSI?

Cylinder Blind End Area = 28.26 square inches
Pressure = 2,500 psi
Pressure × Cylinder Area = 2,500 × 28.26 = 70,650 pounds

What is the pull force of a 6-inch-diameter cylinder with a 3-inch-diameter rod operating at 2,500 PSI?

Cylinder Rod End Area = 21.19 square inches
Pressure = 2,500 psi
Pressure × Cylinder Area = 2,500 × 21.19 = 52,975 pounds

Fluid Pressure in PSI Required to Lift Load (in PSI) = Pounds of Force Needed/Cylinder Area

What pressure is needed to develop 50,000 pounds of push force from a 6-inch-diameter cylinder?

Pounds of Force = 50,000 pounds
Cylinder Blind End Area = 28.26 square inches
Pounds of Force Needed/Cylinder Area = 50,000/28.26 = 1,769.29 psi

What pressure is needed to develop 50,000 pounds of pull force from a 6-inch-diameter cylinder that has a 3-inch-diameter rod?

Pounds of Force = 50,000 pounds
Cylinder Rod End Area = 21.19 square inches
Pounds of Force Needed/Cylinder Area = 50,000/21.19 = 2,359.60 psi

Cylinder Speed (in Inches per Second) = (231 × GPM)/(60 × Net Cylinder Area)

How fast will a 6-inch-diameter cylinder with a 3-inch-diameter rod extend with a 15-gpm input?

GPM = 6

Net Cylinder Area = 28.26 square inches

(231 × GPM)/(60 × Net Cylinder Area) = (231 × 15)/(60 × 28.26) = 2.04 inches per second

How fast will it retract?

Net Cylinder Area = 21.19 square inches

(231 × GPM)/(60 × Net Cylinder Area) = (231 × 15)/(60 × 21.19) = 2.73 inches per second

GPM of Flow Needed for Cylinder Speed = Cylinder Area × Stroke Length in Inches/231 × 60/Time in Seconds for One Stroke

How many GPM are needed to extend a 6-inch-diameter cylinder 8 inches in 10 seconds?

Cylinder Area = 28.26 square inches

Stroke Length = 8 inches

Time for 1 stroke = 10 seconds

Area × Length/231 × 60/Time = 28.26 × 8/231 × 60/10 = 5.88 gpm

If the cylinder has a 3-inch-diameter rod, how many GPM are needed to retract 8 inches in 10 seconds?

Cylinder Area = 21.19 square inches

Stroke Length = 8 inches

Time for 1 stroke = 10 seconds

Area × Length/231 × 60/Time = 21.19 × 8/231 × 60/10 = 4.40 gpm

Cylinder Blind End Output (GPM) = Blind End Area/Rod End Area × GPM In

How many GPM come out the blind end of a 6-inch-diameter cylinder with a 3-inch-diameter rod when there are 15 gallons per minute put in the rod end?

Cylinder Blind End Area = 28.26 square inches

Cylinder Rod End Area = 21.19 square inches

GPM Input = 15 gpm

Blind End Area/Rod End Area × GPM In = 28.26/21.19 × 15 = 20 gpm

Hydraulic Motor Calculations

GPM of Flow Needed for Fluid Motor Speed = Motor Displacement × Motor RPM/231

How many GPM are needed to drive a 2.51 cubic inch motor at 1,200 rpm?

Motor Displacement = 2.51 cubic inches per revolution

Motor RPM = 1,200

Motor Displacement × Motor RPM/231 = 2.51 × 1,200/231 = 13.04 gpm

Fluid Motor Speed from GPM Input = 231 × GPM/Fluid Motor Displacement

How fast will a 0.95 cubic inch motor turn with 8-gpm input?

GPM = 8

Motor Displacement = 0.95 cubic inches per revolution

231 × GPM/Fluid Motor Displacement = 231 × 8/0.95 = 1,945 rpm

Fluid Motor Torque from Pressure and Displacement = PSI × Motor Displacement/(2 × Pi)

How much torque does a 2.25 cubic inch motor develop at 2,200 psi?

Pressure = 2,200 psi

Displacement = 2.25 cubic inches per revolution

PSI × Motor Displacement/(2 × Pi) = 2,200 × 2.25/6.28 = 788.22 inch pounds

Fluid Motor Torque from Horsepower and RPM = Horsepower × 63,025/RPM

How much torque is developed by a motor at 15 horsepower and 1,500 rpm?

Horsepower = 15

RPM = 1,500

Horsepower × 63,025/RPM = 15 × 63,025/1,500 = 630.25 inch pound

Fluid Motor Torque from GPM, PSI, and RPM = GPM × PSI × 36.77/RPM

How much torque does a motor develop at 1,250 psi, 1,750 rpm, with 9-GPM input?

GPM = 9

PSI = 1,250

RPM = 1,750

GPM × PSI × 36.7/RPM = 9 × 1,250 × 36.7/1,750 = 235.93 inch pounds/second

Fluid and Piping Calculations

Velocity of Fluid through Piping = 0.3208 × GPM/Internal Area

What is the velocity of 10 GPM going through a 0.5-inch-diameter schedule 40 pipe?

GPM = 10

Internal Area = 0.304 (see following note)

0.3208 × GPM/Internal Area = 0.3208 × 10 × 0.304 = 10.55 feet per second

Note: The outside diameter of the pipe remains the same regardless of the thickness of the pipe. A heavy duty pipe has a thicker wall than a standard duty pipe, so the internal diameter of the heavy duty pipe is smaller than the internal diameter of a standard duty pipe. The wall thickness and internal diameter of pipes can be found on readily available charts.

Hydraulic steel tubing also maintains the same outside diameter regardless of wall thickness.

Hose sizes indicate the inside diameter of the plumbing. A 0.5-inch-diameter hose has an internal diameter of 0.50 inches, regardless of the hose pressure rating.

SUGGESTED PIPING SIZES

- Pump suction lines should be sized so the fluid velocity is between 2 and 4 feet per second.
- Oil return lines should be sized so the fluid velocity is between 10 and 15 feet per second.
- Medium-pressure supply lines should be sized so the fluid velocity is between 15 and 20 feet per second.
- High-pressure supply lines should be sized so the fluid velocity is below 30 feet per second.

Heat Calculations

Heat Dissipation Capacity of Steel Reservoirs = 0.001 × Surface Area × Difference between Oil and Air Temperature

If the oil temperature is 140 degrees, and the air temperature is 75 degrees, how much heat will a reservoir with 20 square feet of surface area dissipate?

Surface Area = 20 square feet

Temperature Difference = 140 degrees − 75 degrees = 65 degrees

0.001 × Surface Area × Temperature Difference = 0.001 × 20 × 65 = 1.3 horsepower

Note: 1 HP = 2,544 BTU per hour

HEATING HYDRAULIC FLUID

1 watt will raise the temperature of 1 gallon by 1°F per hour
and
horsepower × 745.7 = watts
and
watts/1,000 = kilowatts

APPENDIX

2 Hydraulic/Fluid Power Systems

GENERAL SYSTEM OPERATION

1. Identify system type. Perform system tests and diagnosis.

   The system is:

   ☐ Open center

   ☐ Closed center

   ☐ Closed loop

   ☐ Ask your instructor to assign a system to which you can perform system testing troubleshooting.

   Pressure test results _____

   Flow test results _____

   Diagnosis _____

2. Read and interpret system diagrams and schematics.
   Identify the components by number in the following schematic diagram.

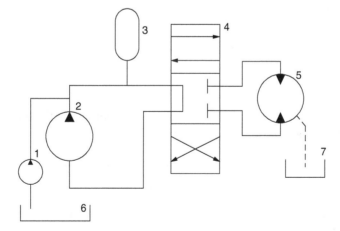

   1. _____
   2. _____
   3. _____
   4. _____
   5. _____
   6. _____
   7. _____

Identify the components on the fluid power system that your instructor assigned to you.

1. _____
2. _____
3. _____
4. _____
5. _____
6. _____
7. _____

3. Perform system temperature, pressure, flow, and cycle time tests. Determine needed actions.

_____
_____
_____
_____
_____
_____

4. Verify placement of equipment/component safety labels and placards. Determine needed actions.

_____
_____
_____
_____
_____
_____

PUMPS

1. Verify proper fluid type.

Use the operating manual for your assigned equipment to determine the type of fluid that should be used in the system. Verify that the oil in the system is the correct type.

The oil that is supposed to be in the system: _____

The oil that is in the system: _____

2. Diagnose causes of pump failure, unusual pump noises, temperatures, flow, and leakage problems. Determine needed actions.

Diagnose a fault in the system you have been assigned:

_____
_____
_____

Describe the action that needs to be taken to resolve the system fault:

_____
_____
_____

© 2011 Cengage Learning. All Rights Reserved. May not be scanned, copied or duplicated, or posted to a publicly accessible website, in whole or in part.

3. Identify pump type, rotation, and drive system. Determine needed actions.

   Pump type:

   _____

   Direction of rotation:

   _____

   What action needs to be taken?

   _____

4. Remove and install pump. Prime and/or bleed system.

   ☐ Completed

5. Inspect pump inlet for restrictions and leaks. Repair as needed.

   ☐ Completed

6. Diagnose root cause of pump failure. Determine needed actions.

   The root cause of the pump failure was:

   _____

   _____

   _____

7. Inspect pump outlet for restrictions and leaks. Repair as needed.

   ☐ Completed

FILTRATION/RESERVOIRS (TANKS)

1. Identify the type of filtration system. Verify filter application and flow direction.

   Type of filtration system: _____

   Specified filter application number: _____

   Actual filter application number: _____

   Direction of flow: _____

2. Service filters and breathers in accordance with manufacturers' recommended procedures.

   ☐ Completed

3. Diagnose cause(s) of system contamination. Determine needed action.

   Cause of contamination:

   _____

   Corrective action:

   _____

4. Take a hydraulic oil sample.

   ☐ Completed

5. Check reservoir fluid level, condition, and consumption. Determine needed action.

   Corrective action:

   _____

6. Inspect and repair or replace the reservoir, sight glass, vents, caps, mounts, valves, screens, and supply and return lines.

☐ Completed

HOSES, FITTINGS, AND CONNECTIONS

1. Diagnose causes of component leakage, damage, and restriction. Determine needed action.

Corrective action:

_____

2. Inspect hoses and connections (length, size, routing, bend radii, and protection). Repair or replace as needed.

☐ Completed

3. Assemble hoses, tubes, connectors, and fittings in accordance with manufacturers' specifications. Use proper procedures to avoid contamination.

☐ Completed

4. Inspect and replace fitting seals and sealants.

☐ Completed

CONTROL VALVES

1. Pressure test system safety relief valve. Determine needed action.

Corrective action:

_____

2. Perform control valve operating pressure and flow tests. Determine needed action.

Corrective action:

_____

3. Inspect, test, and adjust valve controls (electrical/electronic, mechanical, and pneumatic).

☐ Completed

4. Diagnose control valve leakage (internal/external). Determine needed action.

Corrective action:

_____

5. Inspect pilot control valve linkages, cables, and PTO controls. Adjust, repair, or replace as needed.

Corrective action:

_____

ACTUATORS

Actuators comply with manufacturers' and industry-accepted safety practices associated with equipment lock out/tag out, pressure line release, implement/support (blocked or resting on ground), and articulated cylinder devices/machinery safety locks.

1. Identify actuator type (single/double acting, multistage/telescopic, and motors). Determine needed action.

Corrective action:

_____

2. Diagnose the cause of seal failure. Determine needed repairs.

Corrective action:

_____

3. Diagnose the cause of incorrect actuator movement and leakage (internal and external). Determine needed repairs.

Corrective action:

_____

4. Inspect actuator mounting, frame components, and hardware for looseness, cracks, and damage. Repair or replace as needed.

☐ Completed

5. Remove, repair, and/or replace actuators in accordance with manufacturers' recommended procedures.

☐ Completed

6. Inspect actuators for dents, cracks, damage, and leakage. Repair or replace as needed.

☐ Completed

7. Purge and/or bleed system in accordance with manufacturers' recommended procedures.

☐ Completed

Glossary

accumulator In a hydraulic system, this is a device, usually cylindrical in shape and containing a compressible diaphragm or piston, that is used to store fluid either to provide a ready volume or to smooth out pressure pulsations.

acid A liquid or powder that has a deteriorating effect on metal and other materials.

alkaline A liquid or powder that that has a deteriorating effect on organic materials such as skin and rubber compounds.

American National Standards Institute (ANSI) A private, not-for-profit organization that watches over and issues voluntary standards for products, services, processes, systems, and personnel in the United States.

amps A measurement of electrical current flowing through a circuit.

analog An electrical signal where small fluctuations are significant to changing the operation of a system.

analyzing The process of using a compilation of data and information to arrive at a diagnosis or a conclusion about a puzzleor a design.

area A measurement that describes the size of a two-dimensional face of a solid object. It is calculated as a measurement of length, squared. In the case of a rectangle, it is equal to the length times the width. In the case of a circle it is to pi times the square of the radius.

area of a piston A measurement, usually in square inches, that defines the size of the surface of a piston in a hydraulic cylinder.

base *See* alkaline

bidirectional Being able to operate in two, usually opposing, directions, like a fluid power cylinder that can both extend and retract.

Blaise Pascal A 17th-century philosopher and scientist who documented some of the earliest work on pressure's affect on gases and liquids.

Bourdon tube In a gauge a coiled tube that causes the rotation on an arm as it expands and contracts due to changes in pressure applied to the interior of the tube.

Canadian Centre for Occupational Health and Safety A Canadian government organization devoted to ensuring the safety of employees and their workplace.

capacitance The ability of a substance or device to hold an electrical charge.

cavitation The formation of vapor bubbles of a flowing liquid in an area of a system where the pressure of the liquid falls below its vapor pressure.

close center (sometimes referred to as a "closed center") A hydraulic valve that, when moved to the center or idle position, cuts off all fluid flow.

contamination The presence of foreign material, including solids, liquids, and gases into a system or environment.

corrosion The deterioration of any material through a chemical reaction. The most common of these reactions in a fluid power system is the combining of a material with atmospheric oxygen—a process known as oxidation.

cylinder One of the two primary actuators in a fluid power system, the cylinder features a piston that uses force applied by the hydraulic fluid to move a rod to operate the machinery.

data bus A pair of wires designated to carry computer-coded data between electronic control modules.

Delta P Pressure differential.

digital An electronic signal featuring individual, discontinuous voltages to transfer data or to describe a function.

directional control valve A device found in most fluid power systems that controls which direction fluid flows through the actuators, thereby controlling the direction of operation.

entrainment The process by which tiny bubbles of air become trapped in a flowing fluid and are carried throughout the system by that fluid.

extreme high pressure Fluid power systems operating at pressures up to 6,000 psi.

failure mode A general term for operational characteristics of a system when a defect is present.

flats Rotation of a fitting coupler 60 degrees or the equivalent of one side of a hexagonal coupler fitting.

flow meter A test tool designed to measure the mass of fluid flowing through a system in a specified time measurement (i.e., gallons per minute).

flow meter pressure tester This is the basic test instrument for technicians working on fluid power systems and is a combination measurement tool designed to measure not only the flow rate in a system but also the pressure in the system.

flow rate A measurement of the mass of fluid moving through a fluid power system in a specified time measurement, usually expressed in gallons per minute.

force or thrust An influence that is used to accelerate any mass (i.e., move an object).

high pressure Systems with operating pressures ranging from 1,000 to 3,000 psi.

horsepower A unit of measure designed to describe the amount of energy required to move a load over a distance in a specific period of time.

hydraulic cooler A heat exchanger used in a fluid power system to remove heat from the fluid passing thought the system.

hydraulic cylinder *See* cylinder.

hydraulic motor A fluid power system actuator designed to apply a rotary motion or torque to a load or mass.

infrared The part of electromagnetic radiation that is related to heat; the term is often used to describe a device used by technicians to measure the temperature of a component remotely.

inspection An organized examination of a system to determine its state of readiness and operational integrity.

International Standards Organization (ISO) A private, not-for-profit international organization composed of representatives from various national standards organizations that watches over and issues voluntary standards for products, services, processes, systems, and personnel.

interview An information-gathering process whereby a set of specific questions is used to select or eliminate a diagnostic path.

J1587 An on-board computer communication protocol used since the late 1980s to share information

between electronic control units and diagnostic tools on trucks and heavy equipment.

J1939 A high-speed on-board computer communication protocol used since the late 1990s to share information between electronic control units and diagnostic tools on trucks and heavy equipment.

Joint Industrial Council (JIC) A now dormant standards organization formed in the 1950s to establish a uniform set of basic practices for the design and maintenance of machinery.

Joseph Bramah An English inventor and locksmith credited for inventing the first practical hydraulic press. He is considered one of the fathers of hydraulic engineering.

lock-out and tag out A process by which a technician disables the ability of a piece of machinery from being started by installing a lock that can only be removed by that technician; additionally a tag is also added to identify the technician that has disabled the equipment. The technique allows the technician to work on the equipment without the fear of the equipment being started.

lubrication The technique of reducing friction between two contacting surfaces by introducing an intermediary substance called a lubricant.

maintenance intervals An organized calendar of events related to the monitoring, care, and adjustment of a machine.

microprocessor A self-contained integrated circuit capable of receiving and processing data, based on multiple inputs, and issuing a command.

motor (distinguished from a hydraulic motor) An electrical rotational drive device often used as a power supply for hydraulic pumps.

multiplex An electronic communication system employing computer-coded information from multiple sources sharing a common pathway along a pair of wires.

negative temperature coefficient A description of the characteristic of a substance or device that increases in resistance as the temperature decreases.

noncompressible A characteristic of a substance, usually a liquid, the volume of which does not change with the application of pressure.

Occupational Safety and Health Administration (OSHA) A United States government organization devoted to ensuring the safety of employees and their workplace.

ohms The measurement of resistance in an electrical device or conductor.

open center a hydraulic valve that, when moved to the center or idle position, allows fluid to continue to flow through the valve, usually returning it to the reservoir.

potentiometer A variable resistor that serves as an adjustable voltage divider. The output usually varies as a result of physical movement.

power supply A source of power (usually refers to electrical) used to operate a motor, solenoid, sensor, or other device.

pressure A force applied over an area such as pounds per square inch.

pressure compensation Any process in a fluid power system by which the flow rate of the fluid through the system will be altered by changes in pressure.

pressure gauge An analog or digital device used to monitor the pressure in the fluid power system of a machine.

pressure reducing valve A valve or set of valves used to reduce the pressure when a specific set of circumstances or criteria are met.

pressure relief valve A valve placed in a fluid power circuit to bleed off pressure when it reaches a specified level.

pressure sensor A device, usually electronic in nature, used to deliver pressure information to a processor or electronic control unit.

relay An electrical device used to control a large electrical voltage or current flow with a smaller electrical current flow or voltage.

relief valve *See* pressure relief valve.

resistance A property of matter that impedes the ability of electrical current to flow through it.

restriction to flow Any force or obstacle that impedes the flow of fluid through a fluid power system.

rheostat A variable resistor used to change the amount of electrical current flowing through a circuit.

rust Oxidized (corroded) iron.

sequence valve A valve or set of valves used to create a cascading series of actuations based on a progression events.

shuddering An intense low-frequency vibration.

Society of Automotive Engineers (SAE) A professional organization for mobility engineering professionals in the aerospace, automotive, and commercial vehicle industries.

solenoid An electro-magnetic coil device that converts electrical energy into linear motion; usually used to operate a valve or an electrical switch.

solenoid-operated valve An electromagnetic device used to hold and close a valve.

spring return The use of a spring to bring a device to its resting position.

symbol An icon used to represent a specific device or function in a drawing or schematic.

tandem A pair of devices or functions designed to operate together but producing different results.

thermistor A temperature sensitive resistor.

torque Rotational or twisting effect of a force.

travel speed The velocity at which a device such as a cylinder or motor operates.

velocity The distance an item or device can move measured against a specific length of time.

vibration A periodic shaking or oscillation of an object.

volt A measurement of electromotive force, sometimes referred to as the electrical equivalent of pressure.

volume The measurement of matter that describes the three-dimensional space it occupies.

V-Ref (reference voltage) The voltage used to describe the maximum level in an analog electrical circuit—most commonly 5 volts in a modern electronic control system.

watts A measurement of electrical power that is a function of voltage and current flow (amps). A measurement of power equal to 1/746 horsepower.

Index

Page numbers followed by *t* or *f* indicate that the entry is included in a table or figure.

seals
   cylinder leaks of, 178, 178*f*
   high pressure, 30*f*
   hydraulic oils forming, 78–79
   O-ring face, 103
sensors
   angle, 135–136
   boom angle, 131*f*, 136–137, 137*f*
   capacitance, 135, 135*f*
   delta P, 133–134
   flow, 134
   joint angle, 136*f*
   level, 134–135
   load lifting, 135
   load moment, 136*f*
   pressure, 132–133, 133*f*
   temperature, 131–132
sequencing valves, 67, 67*f*, 122*f*
sheath protection, 34*f*
shuddering, 180
shuttle valves, 70, 70*f*
SIDs. *See* Subsystem Identifications
sight glass, 13*f*
simple motor circuits, 119*f*
single acting cylinders, 59–60, 59*f*
skin penetration, 2
Society of Automotive Engineers (SAE), 87,
   103, 139, 141
solenoid-operated control valves, 65, 65*f*, 137,
   137*f*
solenoid-operated valves, 119*f*
spectrographic equipment, 150
split flange O-ring seal fittings, 102–103, 102*f*,
   103*t*
spring-return control valves, 65, 65*f*
stability, of hydraulic fluids, 80
stainless steel tensile strength, 93*t*–94*t*
stall speeds, 164–165
stationary electric motors, 129
steel lines, 90*f*
storage vessel, reservoirs, 40
straight flow, 42
straight thread fitting sizes, 104, 104*t*
straight thread O-ring, 105
strain gauges, 148
Subsystem Identifications (SIDs), 141–142
suction lines, 18*f*
sulfuric acid, 80, 81*f*
supply lines, 14, 17–19, 23*f*
   fluid contamination/velocity in, 19
   fluid flow in, 18–19
   leaks in, 19
   to pumps, 40
   purpose of, 17–18
surface area, of pistons, 7*f*, 32*f*
surface tension, 78, 78*f*
symbols
   ANSI, 57
   control valve modifying, 64–66
   of control valves, 62–64
   of cylinders, 59–60
   of directional control valve, 27*f*

   of filters, 38*f*
   of motors, 60–61
   of power sources, 24*f*
   of pressure relief valves, 24*f*
   of pumps, 20*f*, 58*f*–59*f*
   of reservoirs, 15*f*
   of valves, 66–71
symptoms, 161
synthetic fibers, 153
synthetic fire-resistant fluids, 84

T

technicians, 3–4
telescoping cylinders, 60, 60*f*
temperature
   changes, 18–19
   conversation table, 190
   cylinders changes of, 31–32
   hydraulic fluids influenced by, 80–81
   hydraulic motors changes of, 35
   of hydraulic oils, 163
   inspections and, 155
   operating, 163
   pumps changes of, 22
   reservoirs changes of, 15–16
   sensors, 131–132
thermal efficiency, 21
thermistors, 131–132
thermocouples, 131, 132
threaded seal metal-to-metal straight fittings,
   100, 100*f*
threaded seal metal-to-metal tapered fittings,
   100, 100*f*
three-position detents, 66, 66*f*
three-way selector, 62–63, 62*f*
three-way two-direction control valves, 62, 62*f*
torque
   circular force of, 51*f*
   horsepower relationship with, 51–52, 51*f*
   of hydraulic motors, 50–51, 54,
      115, 116*f*, 116*t*
travel speed, of pistons, 49, 53
troubleshooting
   components and, 169*f*
   contaminants and, 163
   diagnosis critical in, 168*f*
   flow meter testing in, 168
   flow testing in, 167
   fluid flow problems, 169–170
   fluid power systems, 160
   hydraulic systems, 168–169
   information gathering in, 162
   operational testing in, 163–164
   operator interviews in, 160–162
   pressure problems in, 170
   visual inspections in, 145, 162–163, 163*f*
trucks/vehicles, 39*f*, 128*f*, 139
two-way two-position normally closed control
   valves, 62, 62*f*
two-way two-position normally open control
   valves, 62, 62*f*

U

uniform standards, of metric system, 53*f*
units of measurement, 52–53, 187–188
unloading circuits, 123*f*

V

valves. *See also* control valves; directional
   control valves; pressure relief valves
   four-way servo, 70
   latching solenoid-operated, 137
   manual shut-off, 71, 71*f*
   needle, 68
   pressure reducing, 67–68, 67*f*
   relief, 165
   sequencing, 67, 67*f*, 122*f*
   shuttle, 70, 70*f*
   solenoid-operated, 119*f*
   symbols of, 66–71
vane type pumps, 21
vapor, 172–173
vaporization pressure, 173
variable displacement over center pumps, 58, 58*f*
variable displacement piston pumps, 23*f*
variable displacement pressure-compensated
   pumps, 58–59, 59*f*
variable displacement unidirectional motors,
   61, 61*f*
variable displacement unidirectional pumps,
   58, 58*f*
variable orifices, 68, 68*f*
vehicles/trucks, 39*f*, 128*f*, 139
velocity. *See also* fluid velocity
   flow rate/constant, 95*t*–96*t*
   flow rate/diameter interrelationship with, 7–8
   of fluid flow, 49–50, 49*f*, 53
vents, 14*f*, 67
vibration, 180
virtual power, 117*f*
viscosity, of fluids, 79, 79*f*, 83–84, 151
visual inspections, 145, 162–163, 163*f*
voltage, 129, 129*f*
volume
   area/length interrelationship with, 6
   calculating, 6–7
   of cylinders, 7*f*, 48–49, 48*f*, 114*f*
VREF. *See* reference voltage

W

warning devices, 146, 148–149
water
   compressibility of, 54
      glycol systems, 84
   in hydraulic fluids, 81
      oil emulsions, 84
   separation, 82
watts, 130–131, 130*f*
weekly inspections, 146–147
welds, pipe junctions, 146
who/what/where/when questions, 161
wire gauge, 191–192

Printed in the USA
CPSIA information can be obtained
at www.ICGtesting.com
JSHW061113271023
50993JS00004B/53

9 781418 080433